Think DSP
Digital Signal Processing in Python

Allen B. Downey

Beijing · Boston · Farnham · Sebastopol · Tokyo

Think DSP

by Allen B. Downey

Printed in the United States of America.

Published by O'Reilly Media, Inc., 1005 Gravenstein Highway North, Sebastopol, CA 95472.

O'Reilly books may be purchased for educational, business, or sales promotional use. Online editions are also available for most titles (*http://safaribooksonline.com*). For more information, contact our corporate/institutional sales department: 800-998-9938 or *corporate@oreilly.com*.

Editors: Nan Barber and Susan Conant
Production Editor: Kristen Brown
Copyeditor: Kim Cofer
Proofreader: Rachel Head

Indexer: Allen B. Downey
Interior Designer: David Futato
Cover Designer: Karen Montgomery
Illustrator: Rebecca Demarest

July 2016:   First Edition

Revision History for the First Edition
2016-07-11:   First Release

See *http://oreilly.com/catalog/errata.csp?isbn=9781491938454* for release details.

978-1-491-93845-4

[LSI]

Table of Contents

Preface

Signal processing is one of my favorite topics. It is useful in many areas of science and engineering, and if you understand the fundamental ideas, it provides insight into many things we see in the world, and especially the things we hear.

But unless you've studied electrical or mechanical engineering, you probably haven't had a chance to learn about signal processing. The problem is that most books (and the classes that use them) present the material bottom-up, starting with mathematical abstractions like phasors. And they tend to be theoretical, with few applications and little apparent relevance.

The premise of this book is that if you know how to program, you can use that skill to learn other things, and have fun doing it.

With a programming-based approach, I can present the most important ideas right away. By the end of the first chapter, you'll be able to analyze sound recordings and other signals, and generate new sounds. Each chapter introduces a new technique and an application you can apply to real signals. At each step you learn how to use a technique first, and then how it works.

This approach is more practical and, I hope you'll agree, more fun.

Who Is This Book For?

The examples and supporting code for this book are in Python. You should know core Python and you should be familiar with object-oriented features, at least using objects if not defining your own.

If you are not already familiar with Python, you might want to start with my other book, *Think Python*, which is an introduction to Python for people who have never programmed, or Mark Lutz's *Learning Python*, which might be better for people with programming experience.

I use NumPy and SciPy extensively. If you are familiar with them already, that's great, but I will also explain the functions and data structures I use.

I assume that the reader knows basic mathematics, including complex numbers. You don't need much calculus; if you understand the concepts of integration and differentiation, that will do. I use some linear algebra, but I will explain it as we go along.

Using the Code

The code and sound samples used in this book are available from this GitHub repository: *https://github.com/AllenDowney/ThinkDSP*. If you are not familiar with Git and GitHub, Git is a version control system that allows you to keep track of the files that make up a project. A collection of files under Git's control is called a "repository". GitHub is a hosting service that provides storage for Git repositories and a convenient web interface.

The GitHub home page for my repository provides several ways to work with the code:

- You can create a copy of my repository on GitHub by pressing the Fork button. If you don't already have a GitHub account, you'll need to create one. After forking, you'll have your own repository on GitHub that you can use to keep track of code you write while working on this book. Then you can clone the repository, which means that you copy the files to your computer.

- You can clone my repository. You don't need a GitHub account to do this, but you won't be able to write your changes back to GitHub.

- If you don't want to use Git at all, you can download the files in a ZIP file using the button in the lower-right corner of the GitHub page.

All of the code is written to work in both Python 2 and Python 3 with no translation.

I developed this book using Anaconda from Continuum Analytics, which is a free Python distribution that includes all the packages you'll need to run the code (and lots more). I found Anaconda easy to install. By default it does a user-level installation, not system-level, so you don't need administrative privileges. And it supports both Python 2 and Python 3. You can download Anaconda from *http://continuum.io/downloads*.

If you don't want to use Anaconda, you will need the following packages:

- NumPy for basic numerical computation (*http://www.numpy.org*)
- SciPy for scientific computation (*http://www.scipy.org*)
- matplotlib for visualization (*http://matplotlib.org*)

Although these are commonly used packages, they are not included with all Python installations, and they can be hard to install in some environments. If you have trouble installing them, I recommend using Anaconda or one of the other Python distributions that include these packages.

Most exercises use Python scripts, but some also use Jupyter notebooks. If you have not used Jupyter before, you can read about it at *http://jupyter.org*.

There are three ways you can work with the Jupyter notebooks:

Run Jupyter on your computer
> If you installed Anaconda, you probably got Jupyter by default. To check, start the server from the command line, like this:

```
$ jupyter notebook
```

> If it's not installed, you can install it in Anaconda like this:

```
$ conda install jupyter
```

> When you start the server, it should launch your default web browser or create a new tab in an open browser window.

Run Jupyter on Binder
> Binder is a service that runs Jupyter in a virtual machine. If you follow the link *http://mybinder.org/repo/AllenDowney/ThinkDSP*, you should get a Jupyter home page with the notebooks for this book and the supporting data and scripts.

> You can run the scripts and modify them to run your own code, but the virtual machine you run in is temporary. Any changes you make will disappear, along with the virtual machine, if you leave it idle for more than about an hour.

View notebooks on nbviewer
> When I refer to notebooks later in the book, I provide links to nbviewer, which provides a static view of the code and results. You can use these links to read the notebooks and listen to the examples, but you won't be able to modify or run the code, or use the interactive widgets.

Good luck, and have fun!

Conventions Used in This Book

The following typographical conventions are used in this book:

Italic
> Indicates emphasis, keystrokes, menu options, URLs, and email addresses.

Bold
> Used for new terms where they are defined.

Constant width

> Used for program listings, as well as within paragraphs to refer to filenames, file extensions, and program elements such as variable and function names, data types, statements, and keywords.

Constant width bold

> Shows commands or other text that should be typed literally by the user.

Safari® Books Online

 Safari Books Online (*www.safaribooksonline.com*) is an on-demand digital library that delivers expert content in both book and video form from the world's leading authors in technology and business.

Technology professionals, software developers, web designers, and business and creative professionals use Safari Books Online as their primary resource for research, problem solving, learning, and certification training.

Safari Books Online offers a range of plans and pricing for enterprise, government, education, and individuals.

Members have access to thousands of books, training videos, and prepublication manuscripts in one fully searchable database from publishers like O'Reilly Media, Prentice Hall Professional, Addison-Wesley Professional, Microsoft Press, Sams, Que, Peachpit Press, Focal Press, Cisco Press, John Wiley & Sons, Syngress, Morgan Kaufmann, IBM Redbooks, Packt, Adobe Press, FT Press, Apress, Manning, New Riders, McGraw-Hill, Jones & Bartlett, Course Technology, and hundreds more. For more information about Safari Books Online, please visit us online.

How to Contact Us

Please address comments and questions concerning this book to the publisher:

> O'Reilly Media, Inc.
> 1005 Gravenstein Highway North
> Sebastopol, CA 95472
> 800-998-9938 (in the United States or Canada)
> 707-829-0515 (international or local)
> 707-829-0104 (fax)

We have a web page for this book, where we list errata, examples, and any additional information. You can access this page at *http://bit.ly/think-dsp*.

To comment or ask technical questions about this book, send email to *bookquestions@oreilly.com*.

For more information about our books, courses, conferences, and news, see our website at *http://www.oreilly.com*.

Find us on Facebook: *http://facebook.com/oreilly*

Follow us on Twitter: *http://twitter.com/oreillymedia*

Watch us on YouTube: *http://www.youtube.com/oreillymedia*

Contributor List

If you have a suggestion or correction, please send email to *downey@allendowney.com*. If I make a change based on your feedback, I will add you to the contributor list (unless you ask to be omitted).

If you include at least part of the sentence the error appears in, that makes it easy for me to search. Page numbers and section titles are fine, too, but not as easy to work with. Thanks!

- Before I started writing, my thoughts about this book benefited from conversations with Boulos Harb at Google and Aurelio Ramos, formerly at Harmonix Music Systems.

- During the Fall 2013 semester, Nathan Lintz and Ian Daniher worked with me on an independent study project and helped me with the first draft of this book.

- On Reddit's DSP forum, the anonymous user RamjetSoundwave helped me fix a problem with my implementation of Brownian noise. And andodli found a typo.

- In Spring 2015 I had the pleasure of teaching this material along with Prof. Oscar Mur-Miranda and Prof. Siddhartan Govindasamy. Both made many suggestions and corrections.

- Silas Gyger corrected an arithmetic error.

- Giuseppe Masetti sent a number of very helpful suggestions.

Special thanks to the technical reviewers, Eric Peters, Bruce Levens, and John Vincent, for many helpful suggestions, clarifications, and corrections.

Also thanks to Freesound, which is the source of many of the sound samples I use in this book, and to the Freesound users who contributed those samples. I include some of their wave files in the GitHub repository for this book, using the original filenames, so it should be easy to find their sources.

Unfortunately, most Freesound users don't make their real names available, so I can only thank them by their usernames. Samples used in this book were contributed by

Freesound users iluppai, wcfl10, thirsk, docquesting, kleeb, landup, zippi1, themusicalnomad, bcjordan, rockwehrmann, marcgascon7, and jcveliz. Thank you all!

## CHAPTER 1

Sounds and Signals

A **signal** represents a quantity that varies in time. That definition is pretty abstract, so let's start with a concrete example: sound. Sound is variation in air pressure. A sound signal represents variations in air pressure over time.

A microphone is a device that measures these variations and generates an electrical signal that represents sound. A speaker is a device that takes an electrical signal and produces sound. Microphones and speakers are called **transducers** because they transduce, or convert, signals from one form to another.

This book is about signal processing, which includes processes for synthesizing, transforming, and analyzing signals. I will focus on sound signals, but the same methods apply to electronic signals, mechanical vibration, and signals in many other domains.

They also apply to signals that vary in space rather than time, like elevation along a hiking trail. And they apply to signals in more than one dimension, like an image, which you can think of as a signal that varies in two-dimensional space. Or a movie, which is a signal that varies in two-dimensional space *and* time.

But we start with simple one-dimensional sound.

The code for this chapter is in chap01.ipynb, which is in the repository for this book (see "Using the Code" on page viii). You can also view it at *http://tinyurl.com/thinkdsp01*.

Periodic Signals

We'll start with **periodic signals**, which are signals that repeat themselves after some period of time. For example, if you strike a bell, it vibrates and generates sound. If you record that sound and plot the transduced signal, it looks like Figure 1-1.

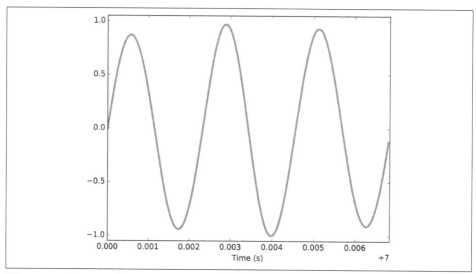

Figure 1-1. Segment from a recording of a bell.

This signal resembles a **sinusoid**, which means it has the same shape as the trigonometric sine function.

You can see that this signal is periodic. I chose the duration to show three full repetitions, also known as **cycles**. The duration of each cycle, called the **period**, is about 2.3 ms.

The **frequency** of a signal is the number of cycles per second, which is the inverse of the period. The units of frequency are cycles per second, or **Hertz**, abbreviated "Hz". (Strictly speaking, the number of cycles is a dimensionless number, so a Hertz is really a "per second".)

The frequency of this signal is about 439 Hz, slightly lower than 440 Hz, which is the standard tuning pitch for orchestral music. The musical name of this note is A, or more specifically, A4. If you are not familiar with "scientific pitch notation", the numerical suffix indicates which octave the note is in. A4 is the A above middle C. A5 is one octave higher. See *http://en.wikipedia.org/wiki/Scientific_pitch_notation*.

A tuning fork generates a sinusoid because the vibration of the tines is a form of simple harmonic motion. Most musical instruments produce periodic signals, but the shape of these signals is not sinusoidal. For example, Figure 1-2 shows a segment from a recording of a violin playing Boccherini's String Quintet No. 5 in E, 3rd movement.

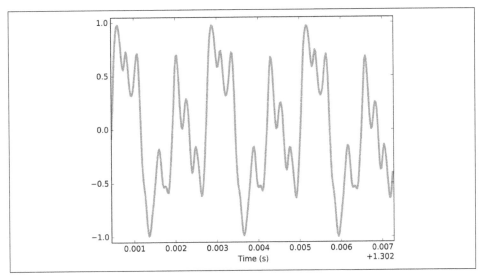

Figure 1-2. Segment from a recording of a violin.

Again we can see that the signal is periodic, but the shape of the signal is more complex. The shape of a periodic signal is called the **waveform**. Most musical instruments produce waveforms more complex than a sinusoid. The shape of the waveform determines the musical **timbre**, which is our perception of the quality of the sound. People usually perceive complex waveforms as rich, warm, and more interesting than sinusoids.

Spectral Decomposition

The most important topic in this book is **spectral decomposition**, which is the idea that any signal can be expressed as the sum of sinusoids with different frequencies.

The most important mathematical idea in this book is the **Discrete Fourier Transform** (DFT), which takes a signal and produces its **spectrum**. The spectrum is the set of sinusoids that add up to produce the signal.

And the most important algorithm in this book is the **Fast Fourier Transform** (FFT), which is an efficient way to compute the DFT.

For example, Figure 1-3 shows the spectrum of the violin recording in Figure 1-2. The x-axis is the range of frequencies that make up the signal. The y-axis shows the strength or **amplitude** of each frequency component.

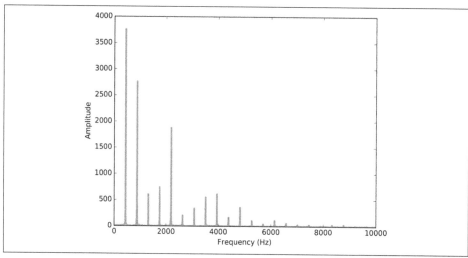

Figure 1-3. Spectrum of a segment from the violin recording.

The lowest frequency component is called the **fundamental frequency**. The fundamental frequency of this signal is near 440 Hz (actually a little lower, or "flat").

In this signal the fundamental frequency has the largest amplitude, so it is also the **dominant frequency**. Normally the perceived pitch of a sound is determined by the fundamental frequency, even if it is not dominant.

The other spikes in the spectrum are at frequencies 880, 1320, 1760, and 2200, which are integer multiples of the fundamental. These components are called **harmonics** because they are musically harmonious with the fundamental:

- 880 is the frequency of A5, one octave higher than the fundamental. An **octave** is a doubling in frequency.

- 1320 is approximately E6, which is a perfect fifth above A5. If you are not familiar with musical intervals like "perfect fifth", see *https://en.wikipedia.org/wiki/Interval_(music)*.

- 1760 is A6, two octaves above the fundamental.

- 2200 is approximately C♯7, which is a major third above A6.

These harmonics make up the notes of an A major chord, although not all in the same octave. Some of them are only approximate because the notes that make up Western music have been adjusted for **equal temperament** (see *http://en.wikipedia.org/wiki/Equal_temperament*).

Given the harmonics and their amplitudes, you can reconstruct the signal by adding up sinusoids. Next we'll see how.

Signals

I wrote a Python module called `thinkdsp.py` that contains classes and functions for working with signals and spectrums[1]. You will find it in the repository for this book (see "Using the Code" on page viii).

To represent signals, `thinkdsp` provides a class called `Signal`. This is the parent class for several signal types, including `Sinusoid`, which represents both sine and cosine signals.

`thinkdsp` provides functions to create sine and cosine signals:

```
cos_sig = thinkdsp.CosSignal(freq=440, amp=1.0, offset=0)
sin_sig = thinkdsp.SinSignal(freq=880, amp=0.5, offset=0)
```

`freq` is frequency in Hz. `amp` is amplitude in unspecified units, where 1.0 is defined as the largest amplitude we can record or play back.

`offset` is a **phase offset** in radians. Phase offset determines where in the period the signal starts. For example, a sine signal with `offset=0` starts at sin 0, which is 0. With `offset=pi/2` it starts at sin $\pi/2$, which is 1.

Signals have an __add__ method, so you can use the + operator to add them:

```
mix = sin_sig + cos_sig
```

The result is a `SumSignal`, which represents the sum of two or more signals.

A `Signal` is basically a Python representation of a mathematical function. Most signals are defined for all values of *t*, from negative infinity to infinity.

You can't do much with a `Signal` until you evaluate it. In this context, "evaluate" means taking a sequence of points in time, `ts`, and computing the corresponding values of the signal, `ys`. I represent `ts` and `ys` using NumPy arrays and encapsulate them in an object called a `Wave`.

A `Wave` represents a signal evaluated at a sequence of points in time. Each point in time is called a **frame** (a term borrowed from movies and video). The measurement itself is called a **sample**, although "frame" and "sample" are sometimes used interchangeably.

`Signal` provides `make_wave`, which returns a new `Wave` object:

```
wave = mix.make_wave(duration=0.5, start=0, framerate=11025)
```

---

1 The plural of "spectrum" is often written "spectra", but I prefer to use standard English plurals. If you are familiar with "spectra", I hope my choice doesn't sound too strange.

duration is the length of the Wave in seconds. start is the start time, also in seconds. framerate is the (integer) number of frames per second, which is also the number of samples per second.

11,025 frames per second is one of several frame rates commonly used in audio file formats, including Waveform Audio File (WAV) and MP3.

This example evaluates the signal from $t=0$ to $t=0.5$ at 5513 equally spaced frames (because 5513 is half of 11,025). The time between frames, or **timestep**, is 1/1,1025 seconds, about 91 μs.

Wave provides a plot method that uses pyplot. You can plot the wave like this:

```
wave.plot()
pyplot.show()
```

pyplot is part of matplotlib; it is included in many Python distributions, or you might have to install it.

At freq=440 there are 220 periods in 0.5 seconds, so this plot would look like a solid block of color. To zoom in on a small number of periods, we can use segment, which copies a segment of a Wave and returns a new wave:

```
period = mix.period
segment = wave.segment(start=0, duration=period*3)
```

period is a property of a Signal; it returns the period in seconds.

start and duration are in seconds. This example copies the first three periods from mix. The result is a Wave object.

If we plot segment, it looks like Figure 1-4. This signal contains two frequency components, so it is more complicated than the signal from the tuning fork, but less complicated than the violin.

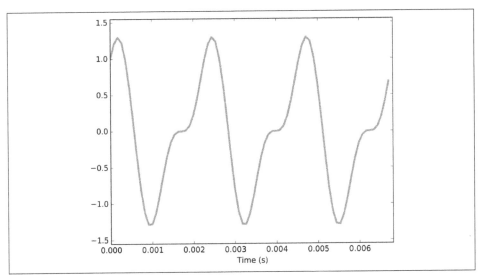

Figure 1-4. Segment from a mixture of two sinusoid signals.

Reading and Writing Waves

thinkdsp provides read_wave, which reads a WAV file and returns a Wave:

```
violin_wave = thinkdsp.read_wave('input.wav')
```

And Wave provides write, which writes a WAV file:

```
wave.write(filename='output.wav')
```

You can listen to the Wave with any media player that plays WAV files. On Unix systems I use aplay, which is simple, robust, and included in many Linux distributions.

thinkdsp also provides play_wave, which runs the media player as a subprocess:

```
thinkdsp.play_wave(filename='output.wav', player='aplay')
```

It uses aplay by default, but you can provide the name of another player.

Spectrums

Wave provides make_spectrum, which returns a Spectrum:

```
spectrum = wave.make_spectrum()
```

And Spectrum provides plot:

```
spectrum.plot()
thinkplot.show()
```

`thinkplot` is a module I wrote to provide wrappers around some of the functions in `pyplot`. It is included in the Git repository for this book (see "Using the Code" on page viii).

`Spectrum` provides three methods that modify the spectrum:

- `low_pass` applies a low-pass filter, which means that components above a given cutoff frequency are **attenuated** (that is, reduced in magnitude) by a factor.

- `high_pass` applies a high-pass filter, which means that it attenuates components below the cutoff.

- `band_stop` attenuates components in the band of frequencies between two cutoffs.

This example attenuates all frequencies above 600 by 99%:

```
spectrum.low_pass(cutoff=600, factor=0.01)
```

A low-pass filter removes bright, high-frequency sounds, so the result sounds muffled and darker. To hear what it sounds like, you can convert the `Spectrum` back to a `Wave`, and then play it:

```
wave = spectrum.make_wave()
wave.play('temp.wav')
```

The `play` method writes the wave to a file and then plays it. If you use Jupyter notebooks, you can use `make_audio`, which makes an Audio widget that plays the sound.

Wave Objects

There is nothing very complicated in `thinkdsp.py`. Most of the functions it provides are thin wrappers around functions from NumPy and SciPy.

The primary classes in `thinkdsp` are `Signal`, `Wave`, and `Spectrum`. Given a `Signal`, you can make a `Wave`. Given a `Wave`, you can make a `Spectrum`, and vice versa. These relationships are shown in Figure 1-5.

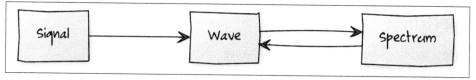

Figure 1-5. Relationships among the classes in `thinkdsp`.

A `Wave` object contains three attributes: `ys` is a NumPy array that contains the values in the signal; `ts` is an array of the times where the signal was evaluated or sampled;

and `framerate` is the number of samples per unit of time. The unit of time is usually seconds, but it doesn't have to be. In one of my examples, it's days.

`Wave` also provides three read-only properties: `start`, `end`, and `duration`. If you modify `ts`, these properties change accordingly.

To modify a wave, you can access the `ts` and `ys` directly. For example:

```
wave.ys *= 2
wave.ts += 1
```

The first line scales the wave by a factor of 2, making it louder. The second line shifts the wave in time, making it start 1 second later.

But `Wave` provides methods that perform many common operations. For example, the same two transformations could be written:

```
wave.scale(2)
wave.shift(1)
```

You can read the documentation of these methods and others at *http://greentea press.com/thinkdsp.html*.

Signal Objects

`Signal` is a parent class that provides functions common to all kinds of signals, like `make_wave`. Child classes inherit these methods and provide `evaluate`, which evaluates the signal at a given sequence of times.

For example, `Sinusoid` is a child class of `Signal`, with this definition:

```
class Sinusoid(Signal):

    def __init__(self, freq=440, amp=1.0, offset=0, func=np.sin):
        Signal.__init__(self)
        self.freq = freq
        self.amp = amp
        self.offset = offset
        self.func = func
```

The parameters of `__init__` are:

`freq`
  Frequency in cycles per second, or Hz.

`amp`
  Amplitude. The units of amplitude are arbitrary, usually chosen so 1.0 corresponds to the maximum input from a microphone or maximum output to a speaker.

offset

Indicates where in its period the signal starts; `offset` is in units of radians.

func

A Python function used to evaluate the signal at a particular point in time. It is usually either `np.sin` or `np.cos`, yielding a sine or cosine signal.

Like many init methods, this one just tucks the parameters away for future use.

`Signal` provides `make_wave`, which looks like this:

```
def make_wave(self, duration=1, start=0, framerate=11025):
    n = round(duration * framerate)
    ts = start + np.arange(n) / framerate
    ys = self.evaluate(ts)
    return Wave(ys, ts, framerate=framerate)
```

`start` and `duration` are the start time and duration in seconds. `framerate` is the number of frames (samples) per second.

`n` is the number of samples, and `ts` is a NumPy array of sample times.

To compute the `ys`, `make_wave` invokes `evaluate`, which is provided by `Sinusoid`:

```
def evaluate(self, ts):
    phases = PI2 * self.freq * ts + self.offset
    ys = self.amp * self.func(phases)
    return ys
```

Let's unwind this function one step at a time:

1. `self.freq` is frequency in cycles per second, and each element of `ts` is a time in seconds, so their product is the number of cycles since the start time.

2. `PI2` is a constant that stores 2π. Multiplying by `PI2` converts from cycles to **phase**. You can think of phase as "cycles since the start time" expressed in radians. Each cycle is 2π radians.

3. `self.offset` is the phase when $t = 0$. It has the effect of shifting the signal left or right in time.

4. If `self.func` is `np.sin` or `np.cos`, the result is a value between -1 and $+1$.

5. Multiplying by `self.amp` yields a signal that ranges from `-self.amp` to `+self.amp`.

In math notation, `evaluate` is written like this:

$$y = A \cos \left(2\pi f t + \phi_0 \right)$$

where A is amplitude, f is frequency, t is time, and ϕ_0 is the phase offset. It may seem like I wrote a lot of code to evaluate one simple expression, but as we'll see, this code provides a framework for dealing with all kinds of signals, not just sinusoids.

Exercises

Before you begin these exercises, you should download the code for this book, following the instructions in "Using the Code" on page viii.

Solutions to these exercises are in `chap01soln.ipynb`.

Exercise 1-1.

If you have Jupyter, load `chap01.ipynb`, read through it, and run the examples. You can also view this notebook at *http://tinyurl.com/thinkdsp01*.

Exercise 1-2.

Go to *http://freesound.org* and download a sound sample that includes music, speech, or other sounds that have a well-defined pitch. Select a roughly half-second segment where the pitch is constant. Compute and plot the spectrum of the segment you selected. What connection can you make between the timbre of the sound and the harmonic structure you see in the spectrum?

Use `high_pass`, `low_pass`, and `band_stop` to filter out some of the harmonics. Then convert the spectrum back to a wave and listen to it. How does the sound relate to the changes you made in the spectrum?

Exercise 1-3.

Synthesize a compound signal by creating `SinSignal` and `CosSignal` objects and adding them up. Evaluate the signal to get a `Wave`, and listen to it. Compute its `Spec trum` and plot it. What happens if you add frequency components that are not multiples of the fundamental?

Exercise 1-4.

Write a function called `stretch` that takes a `Wave` and a stretch factor and speeds up or slows down the wave by modifying `ts` and `framerate`. Hint: it should only take two lines of code.

CHAPTER 2

Harmonics

In this chapter I present several new waveforms; we will look at their spectrums to understand their **harmonic structure**, which is the set of sinusoids they are made up of.

I'll also introduce one of the most important phenomena in digital signal processing: aliasing. And I'll explain a little more about how the Spectrum class works.

The code for this chapter is in chap02.ipynb, which is in the repository for this book (see "Using the Code" on page viii). You can also view it at *http://tinyurl.com/ thinkdsp02*.

Triangle Waves

A sinusoid contains only one frequency component, so its spectrum has only one peak. More complicated waveforms, like the violin recording in Figure 1-2, yield DFTs with many peaks. In this section we investigate the relationship between waveforms and their spectrums.

I'll start with a triangle waveform, which is like a straight-line version of a sinusoid. Figure 2-1 shows a triangle waveform with frequency 200 Hz.

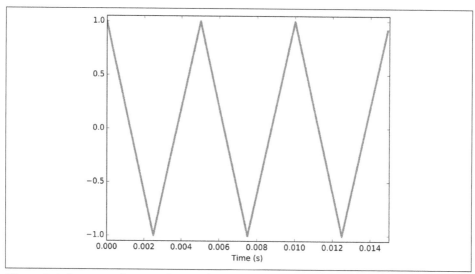

Figure 2-1. Segment of a triangle signal at 200 Hz.

To generate a triangle wave, you can start with a `thinkdsp.TriangleSignal`:

```
class TriangleSignal(Sinusoid):

    def evaluate(self, ts):
        cycles = self.freq * ts + self.offset / PI2
        frac, _ = np.modf(cycles)
        ys = np.abs(frac - 0.5)
        ys = normalize(unbias(ys), self.amp)
        return ys
```

`TriangleSignal` inherits `__init__` from `Sinusoid`, so it takes the same arguments: `freq`, `amp`, and `offset`.

The only difference is `evaluate`. As we saw before, `ts` is the sequence of sample times where we want to evaluate the signal.

There are many ways to generate a triangle wave. The details are not important, but here's how `evaluate` works:

1. `cycles` is the number of cycles since the start time. `np.modf` splits the number of cycles into the fraction part, stored in `frac`, and the integer part, which is ignored.[1]

---

1 Using an underscore as a variable name is a convention that means, "I don't intend to use this value."

2. `frac` is a sequence that ramps from 0 to 1 with the given frequency. Subtracting 0.5 yields values between –0.5 and 0.5. Taking the absolute value yields a waveform that zigzags between 0.5 and 0.

3. `unbias` shifts the waveform down so it is centered at 0; then `normalize` scales it to the given amplitude, `amp`.

Here's the code that generates Figure 2-1:

```
signal = thinkdsp.TriangleSignal(200)
signal.plot()
```

Next we can use the `Signal` to make a `Wave`, and use the `Wave` to make a `Spectrum`:

```
wave = signal.make_wave(duration=0.5, framerate=10000)
spectrum = wave.make_spectrum()
spectrum.plot()
```

Figure 2-2 shows two views of the result; the view on the right is scaled to show the harmonics more clearly. As expected, the highest peak is at the fundamental frequency, 200 Hz, and there are additional peaks at harmonic frequencies, which are integer multiples of 200.

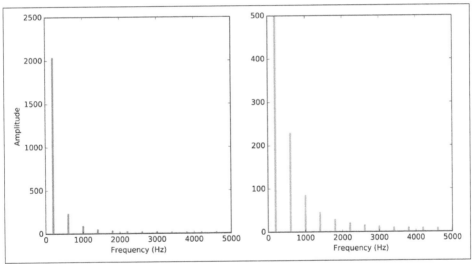

Figure 2-2. Spectrum of a triangle signal at 200 Hz, shown on two vertical scales. The version on the right cuts off the fundamental to show the harmonics more clearly.

But one surprise is that there are no peaks at the even multiples: 400, 800, etc. The harmonics of a triangle wave are all odd multiples of the fundamental frequency, in this example 600, 1000, 1400, etc.

Another feature of this spectrum is the relationship between the amplitude and frequency of the harmonics. Their amplitude drops off in proportion to frequency squared. For example, the frequency ratio of the first two harmonics (200 and 600 Hz) is 3, and the amplitude ratio is approximately 9. The frequency ratio of the next two harmonics (600 and 1000 Hz) is 1.7, and the amplitude ratio is approximately $1.7^2 = 2.9$. This relationship is called the **harmonic structure**.

Square Waves

thinkdsp also provides SquareSignal, which represents a square signal. Here's the class definition:

```
class SquareSignal(Sinusoid):

    def evaluate(self, ts):
        cycles = self.freq * ts + self.offset / PI2
        frac, _ = np.modf(cycles)
        ys = self.amp * np.sign(unbias(frac))
        return ys
```

Like TriangleSignal, SquareSignal inherits __init__ from Sinusoid, so it takes the same parameters.

And the evaluate method is similar. Again, cycles is the number of cycles since the start time, and frac is the fractional part, which ramps from 0 to 1 each period.

unbias shifts frac so it ramps from –0.5 to 0.5, then np.sign maps the negative values to –1 and the positive values to 1. Multiplying by amp yields a square wave that jumps between -amp and amp.

Figure 2-3 shows three periods of a square wave with frequency 100 Hz, and Figure 2-4 shows its spectrum.

Like a triangle wave, the square wave contains only odd harmonics, which is why there are peaks at 300, 500, and 700 Hz, etc. But the amplitude of the harmonics drops off more slowly. Specifically, amplitude drops in proportion to frequency (not frequency squared).

The exercises at the end of this chapter give you a chance to explore other waveforms and other harmonic structures.

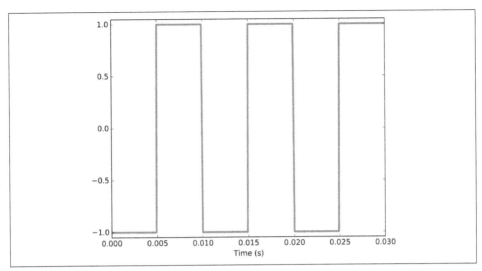

Figure 2-3. Segment of a square signal at 100 Hz.

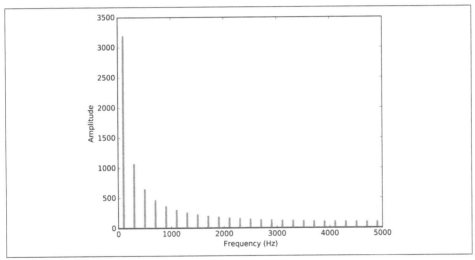

Figure 2-4. Spectrum of a square signal at 100 Hz.

Aliasing

I have a confession. I chose the examples in the previous section carefully to avoid showing you something confusing. But now it's time to get confused.

Figure 2-5 shows the spectrum of a triangle wave at 1100 Hz, sampled at 10,000 frames per second. Again, the view on the right is scaled to show the harmonics.

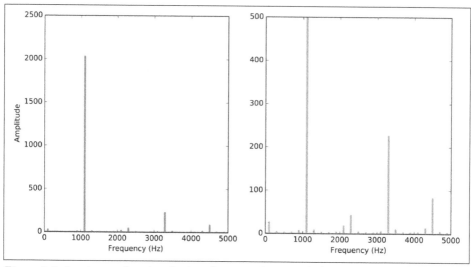

Figure 2-5. Spectrum of a triangle signal at 1100 Hz sampled at 10,000 frames per second. The view on the right is scaled to show the harmonics.

The harmonics of this wave should be at 3300, 5500, 7700, and 9900 Hz. In the figure, there are peaks at 1100 and 3300 Hz, as expected, but the third peak is at 4500, not 5500 Hz. The fourth peak is at 2300, not 7700 Hz. And if you look closely, the peak that should be at 9900 is actually at 100 Hz. What's going on?

The problem is that when you evaluate the signal at discrete points in time, you lose information about what happened between samples. For low-frequency components, that's not a problem, because you have lots of samples per period.

But if you sample a signal at 5000 Hz with 10,000 frames per second, you only have two samples per period. That turns out to be enough, just barely, but if the frequency is higher, it's not.

To see why, let's generate cosine signals at 4500 and 5500 Hz, and sample them at 10,000 frames per second:

```
framerate = 10000

signal = thinkdsp.CosSignal(4500)
duration = signal.period*5
segment = signal.make_wave(duration, framerate=framerate)
segment.plot()

signal = thinkdsp.CosSignal(5500)
segment = signal.make_wave(duration, framerate=framerate)
segment.plot()
```

Figure 2-6 shows the result. I plotted the Signals with thin gray lines and the samples using vertical lines, to make it easier to compare the two Waves. The problem should be clear: even though the Signals are different, the Waves are identical!

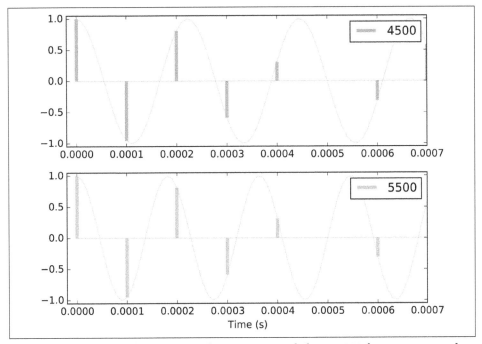

Figure 2-6. Cosine signals at 4500 and 5500 Hz, sampled at 10,000 frames per second. The signals are different, but the samples are identical.

When we sample a 5500 Hz signal at 10,000 frames per second, the result is indistinguishable from a 4500 Hz signal. For the same reason, a 7700 Hz signal is indistinguishable from 2300 Hz, and a 9900 Hz signal is indistinguishable from 100 Hz.

This effect is called **aliasing** because when the high-frequency signal is sampled, it appears to be a low-frequency signal.

In this example, the highest frequency we can measure is 5000 Hz, which is half the sampling rate. Frequencies above 5000 Hz are folded back below 5000 Hz, which is why this threshold is sometimes called the "folding frequency". It is sometimes also called the **Nyquist frequency**. See *http://en.wikipedia.org/wiki/Nyquist_frequency*.

The folding pattern continues if the aliased frequency goes below zero. For example, the fifth harmonic of the 1100 Hz triangle wave is at 12,100 Hz. Folded at 5000 Hz, it would appear at –2100 Hz, but it gets folded again at 0 Hz, back to 2100 Hz. In fact, you can see a small peak at 2100 Hz in Figure 2-4, and the next one at 4300 Hz.

Computing the Spectrum

We have seen the Wave method make_spectrum several times. Here is the implementation (leaving out some details we'll get to later):

```
from np.fft import rfft, rfftfreq

# class Wave:
    def make_spectrum(self):
        n = len(self.ys)
        d = 1 / self.framerate

        hs = rfft(self.ys)
        fs = rfftfreq(n, d)

        return Spectrum(hs, fs, self.framerate)
```

The parameter self is a Wave object. n is the number of samples in the wave, and d is the inverse of the frame rate, which is the time between samples.

np.fft is the NumPy module that provides functions related to the Fast Fourier Transform (FFT), which is an efficient algorithm that computes the Discrete Fourier Transform (DFT).

make_spectrum uses rfft, which stands for "real FFT", because the Wave contains real values, not complex. Later we'll see the full FFT, which can handle complex signals (see "DFT of Real Signals" on page 88). The result of rfft, which I call hs, is a NumPy array of complex numbers that represents the amplitude and phase offset of each frequency component in the wave.

The result of rfftfreq, which I call fs, is an array that contains frequencies corresponding to the hs.

To understand the values in hs, consider these two ways to think about complex numbers:

- A complex number is the sum of a real part and an imaginary part, often written $x + iy$, where i is the imaginary unit $\sqrt{-1}$. You can think of x and y as Cartesian coordinates.

- A complex number is also the product of a magnitude and a complex exponential, $Ae^{i\phi}$, where A is the **magnitude** and ϕ is the **angle** in radians, also called the "argument". You can think of A and ϕ as polar coordinates.

Each value in hs corresponds to a frequency component: its magnitude is proportional to the amplitude of the corresponding component; its angle is the phase offset.

The Spectrum class provides two read-only properties, amps and angles, which return NumPy arrays representing the magnitudes and angles of the hs. When we plot a `Spectrum` object, we usually plot `amps` versus `fs`. Sometimes it is also useful to plot `angles` versus `fs`.

Although it might be tempting to look at the real and imaginary parts of `hs`, you will almost never need to. I encourage you to think of the DFT as a vector of amplitudes and phase offsets that happen to be encoded in the form of complex numbers.

To modify a `Spectrum`, you can access the `hs` directly. For example:

```
spectrum.hs *= 2
spectrum.hs[spectrum.fs > cutoff] = 0
```

The first line multiplies the elements of `hs` by 2, which doubles the amplitudes of all components. The second line sets to 0 only the elements of `hs` where the corresponding frequency exceeds some cutoff frequency.

But `Spectrum` also provides methods to perform these operations:

```
spectrum.scale(2)
spectrum.low_pass(cutoff)
```

You can read the documentation of these methods and others at *http://greentea press.com/thinkdsp.html*.

At this point you should have a better idea of how the `Signal`, `Wave`, and `Spectrum` classes work, but I have not explained how the Fast Fourier Transform works. That will take a few more chapters.

Exercises

Solutions to these exercises are in `chap02soln.ipynb`.

Exercise 2-1.

If you use Jupyter, load `chap02.ipynb` and try out the examples. You can also view the notebook at *http://tinyurl.com/thinkdsp02*.

Exercise 2-2.

A sawtooth signal has a waveform that ramps up linearly from −1 to 1, then drops to −1 and repeats. See *http://en.wikipedia.org/wiki/Sawtooth_wave*.

Write a class called `SawtoothSignal` that extends `Signal` and provides `evaluate` to evaluate a sawtooth signal.

Compute the spectrum of a sawtooth wave. How does the harmonic structure compare to triangle and square waves?

Exercise 2-3.

Make a square signal at 1100 Hz and make a Wave that samples it at 10,000 frames per second. If you plot the spectrum, you can see that most of the harmonics are aliased. When you listen to the wave, can you hear the aliased harmonics?

Exercise 2-4.

If you have a spectrum object, spectrum, and print the first few values of spec trum.fs, you'll see that they start at zero. So spectrum.hs[0] is the magnitude of the component with frequency 0. But what does that mean?

Try this experiment:

1. Make a triangle signal with frequency 440 and make a Wave with duration 0.01 seconds. Plot the waveform.

2. Make a Spectrum object and print spectrum.hs[0]. What is the amplitude and phase of this component?

3. Set spectrum.hs[0] = 100. What effect does this operation have on the waveform? Hint: Spectrum provides a method called make_wave that computes the Wave that corresponds to the Spectrum.

Exercise 2-5.

Write a function that takes a Spectrum as a parameter and modifies it by dividing each element of hs by the corresponding frequency from fs. Hint: since division by zero is undefined, you might want to set spectrum.hs[0] = 0.

Test your function using a square, triangle, or sawtooth wave:

1. Compute the Spectrum and plot it.

2. Modify the Spectrum using your function and plot it again.

3. Use Spectrum.make_wave to make a Wave from the modified Spectrum, and listen to it. What effect does this operation have on the signal?

Exercise 2-6.

Triangle and square waves have odd harmonics only; the sawtooth wave has both even and odd harmonics. The harmonics of the square and sawtooth waves drop off in proportion to $1/f$; the harmonics of the triangle wave drop off like $1/f^2$. Can you find a waveform that has even and odd harmonics that drop off like $1/f^2$?

Hint: there are two ways you could approach this. You could construct the signal you want by adding up sinusoids, or you could start with a signal that is similar to what you want and modify it.

## CHAPTER 3

Non-Periodic Signals

The signals we have worked with so far are periodic, which means that they repeat forever. It also means that the frequency components they contain do not change over time. In this chapter, we consider non-periodic signals, whose frequency components *do* change over time. In other words, pretty much all sound signals.

This chapter also presents spectrograms, a common way to visualize non-periodic signals.

The code for this chapter is in `chap03.ipynb`, which is in the repository for this book (see "Using the Code" on page viii). You can also view it at *http://tinyurl.com/ thinkdsp03*.

Linear Chirp

We'll start with a **chirp**, which is a signal with variable frequency. `thinkdsp` provides a `Signal` called `Chirp` that makes a sinusoid that sweeps linearly through a range of frequencies.

Here's an example that sweeps from 220 to 880 Hz, which is two octaves from A3 to A5:

```
signal = thinkdsp.Chirp(start=220, end=880)
wave = signal.make_wave()
```

Figure 3-1 shows segments of this wave near the beginning, middle, and end. It's clear that the frequency is increasing.

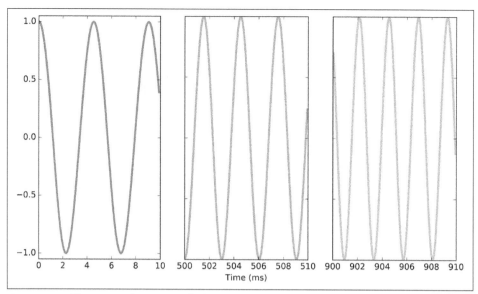

Figure 3-1. Chirp waveform near the beginning, middle, and end.

Before we go on, let's see how `Chirp` is implemented. Here is the class definition:

```
class Chirp(Signal):

    def __init__(self, start=440, end=880, amp=1.0):
        self.start = start
        self.end = end
        self.amp = amp
```

`start` and `end` are the frequencies, in Hz, at the start and end of the chirp. `amp` is amplitude.

Here is the function that evaluates the signal:

```
def evaluate(self, ts):
    freqs = np.linspace(self.start, self.end, len(ts)-1)
    return self._evaluate(ts, freqs)
```

`ts` is the sequence of points in time where the signal should be evaluated; to keep this function simple, I assume they are equally spaced.

If the length of `ts` is n, you can think of it as a sequence of $n - 1$ intervals of time. To compute the frequency during each interval, I use `np.linspace`, which returns a NumPy array of $n - 1$ values between `start` and `end`.

`_evaluate` is a private method that does the rest of the math[1]:

```
def _evaluate(self, ts, freqs):
    dts = np.diff(ts)
    dphis = PI2 * freqs * dts
    phases = np.cumsum(dphis)
    phases = np.insert(phases, 0, 0)
    ys = self.amp * np.cos(phases)
    return ys
```

`np.diff` computes the difference between adjacent elements of `ts`, returning the length of each interval in seconds. If the elements of `ts` are equally spaced, the `dts` values are all the same.

The next step is to figure out how much the phase changes during each interval. In "Signal Objects" on page 9 we saw that when frequency is constant, the phase, ϕ, increases linearly over time:

$$\phi = 2\pi f t$$

When frequency is a function of time, the *change* in phase during a short time interval, Δt, is:

$$\Delta\phi = 2\pi f(t)\Delta t$$

In Python, since `freqs` contains $f(t)$ and `dts` contains the time intervals, we can write:

```
dphis = PI2 * freqs * dts
```

Now, since `dphis` contains the changes in phase, we can get the total phase at each timestep by adding up the changes:

```
phases = np.cumsum(dphis)
phases = np.insert(phases, 0, 0)
```

`np.cumsum` computes the cumulative sum, which is almost what we want, but it doesn't start at 0. So I use `np.insert` to add a 0 at the beginning.

The result is a NumPy array where the `i`th element contains the sum of the first `i` terms from `dphis`; that is, the total phase at the end of the `i`th interval. Finally, `np.cos` computes the amplitude of the wave as a function of phase (remember that phase is expressed in radians).

---

1 Beginning a method name with an underscore makes it "private", indicating that it is not part of the API that should be used outside the class definition.

If you know calculus, you might notice that the limit as Δt gets small is:

$$d\phi = 2\pi f(t)dt$$

Dividing through by dt yields:

$$\frac{d\phi}{dt} = 2\pi f(t)$$

In other words, frequency is the derivative of phase. Conversely, phase is the integral of frequency. When we used cumsum to go from frequency to phase, we were approximating integration.

Exponential Chirp

When you listen to this chirp, you might notice that the pitch rises quickly at first and then slows down. The chirp spans two octaves, but it only takes 2/3 s to span the first octave, and twice as long to span the second.

The reason is that our perception of pitch depends on the logarithm of frequency. As a result, the **interval** we hear between two notes depends on the *ratio* of their frequencies, not the difference. "Interval" is the musical term for the perceived difference between two pitches.

For example, an octave is an interval where the ratio of two pitches is 2. So the interval from 220 to 440 Hz is one octave and the interval from 440 to 880 Hz is also one octave. The difference in frequency is bigger, but the ratio is the same.

As a result, if frequency increases linearly, as in a linear chirp, the perceived pitch increases logarithmically.

If you want the perceived pitch to increase linearly, the frequency has to increase exponentially. A signal with that shape is called an **exponential chirp**.

Here's the ExpoChirp class definition:

```
class ExpoChirp(Chirp):

    def evaluate(self, ts):
        start, end = np.log10(self.start), np.log10(self.end)
        freqs = np.logspace(start, end, len(ts)-1)
        return self._evaluate(ts, freqs)
```

Instead of np.linspace, this version of evaluate uses np.logspace, which creates a series of frequencies whose logarithms are equally spaced, which means that they increase exponentially.

That's it; everything else is the same as `Chirp`. Here's the code that makes one:

```
signal = thinkdsp.ExpoChirp(start=220, end=880)
wave = signal.make_wave(duration=1)
```

You can listen to the examples in `chap03.ipynb` and compare the linear and exponential chirps.

Spectrum of a Chirp

What do you think happens if you compute the spectrum of a chirp? Here's an example that constructs a one-second, one-octave chirp and its spectrum:

```
signal = thinkdsp.Chirp(start=220, end=440)
wave = signal.make_wave(duration=1)
spectrum = wave.make_spectrum()
```

Figure 3-2 shows the result. The spectrum has components at every frequency from 220 to 440 Hz, with variations that look a little like the Eye of Sauron (see *http://en.wikipedia.org/wiki/Sauron*).

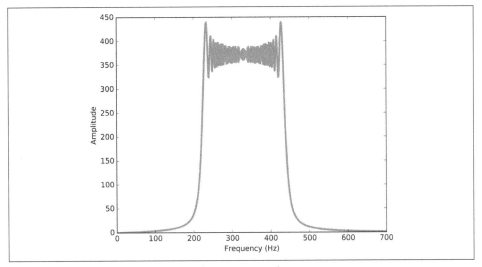

Figure 3-2. Spectrum of a one-second, one-octave chirp.

The spectrum is approximately flat between 220 and 440 Hz, which indicates that the signal spends equal time at each frequency in this range. Based on that observation, you should be able to guess what the spectrum of an exponential chirp looks like.

The spectrum gives hints about the structure of the signal, but it obscures the relationship between frequency and time. For example, we cannot tell by looking at this spectrum whether the frequency went up or down, or both.

Spectrogram

To recover the relationship between frequency and time, we can break the chirp into segments and plot the spectrum of each segment. The result is called a **Short-Time Fourier Transform** (STFT).

There are several ways to visualize an STFT, but the most common is a **spectrogram**, which shows time on the x-axis and frequency on the y-axis. Each column in the spectrogram shows the spectrum of a short segment, using color or grayscale to represent amplitude.

As an example, I'll compute the spectrogram of this chirp:

```
signal = thinkdsp.Chirp(start=220, end=440)
wave = signal.make_wave(duration=1, framerate=11025)
```

Wave provides `make_spectrogram`, which returns a `Spectrogram` object:

```
spectrogram = wave.make_spectrogram(seg_length=512)
spectrogram.plot(high=700)
```

`seg_length` is the number of samples in each segment. I chose 512 because the FFT is most efficient when the number of samples is a power of 2.

Figure 3-3 shows the result. The x-axis shows time from 0 to 1 seconds. The y-axis shows frequency from 0 to 700 Hz. I cut off the top part of the spectrogram; the full range goes to 5512.5 Hz, which is half of the frame rate.

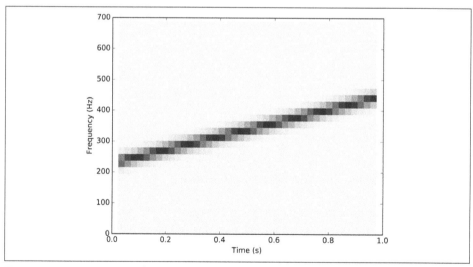

Figure 3-3. Spectrogram of a one-second, one-octave chirp.

The spectrogram shows clearly that frequency increases linearly over time. However, notice that the peak in each column is blurred across 2–3 cells. This blurring reflects the limited resolution of the spectrogram.

The Gabor Limit

The **time resolution** of the spectrogram is the duration of the segments, which corresponds to the width of the cells in the spectrogram. Since each segment is 512 frames, and there are 11,025 frames per second, the duration of each segment is about 0.046 seconds.

The **frequency resolution** is the frequency range between elements in the spectrum, which corresponds to the height of the cells. With 512 frames, we get 256 frequency components over a range from 0 to 5512.5 Hz, so the range between components is 21.6 Hz.

More generally, if n is the segment length, the spectrum contains $n/2$ components. If the frame rate is r, the maximum frequency in the spectrum is $r/2$. So the time resolution is n/r and the frequency resolution is:

$$\frac{r/2}{n/2}$$

which is r/n.

Ideally we would like time resolution to be small, so we can see rapid changes in frequency. And we would like frequency resolution to be small so we can see small changes in frequency. But you can't have both. Notice that time resolution, n/r, is the inverse of frequency resolution, r/n. So if one gets smaller, the other gets bigger.

For example, if you double the segment length, you cut frequency resolution in half (which is good), but you double time resolution (which is bad). Even increasing the frame rate doesn't help. You get more samples, but the range of frequencies increases at the same time.

This tradeoff is called the **Gabor limit** and it is a fundamental limitation of this kind of time–frequency analysis.

Leakage

In order to explain how make_spectrogram works, I have to explain windowing; and in order to explain windowing, I have to show you the problem it is meant to address, which is leakage.

The Discrete Fourier Transform (DFT), which we use to compute Spectrums, treats waves as if they are periodic; that is, it assumes that the finite segment it operates on is a complete period from an infinite signal that repeats over all time. In practice, this assumption is often false, which creates problems.

One common problem is discontinuity at the beginning and end of the segment. Because the DFT assumes that the signal is periodic, it implicitly connects the end of the segment back to the beginning to make a loop. If the end does not connect smoothly to the beginning, the discontinuity creates additional frequency components in the segment that are not in the signal.

As an example, let's start with a sine signal that contains only one frequency component at 440 Hz:

```
signal = thinkdsp.SinSignal(freq=440)
```

If we select a segment that happens to be an integer multiple of the period, the end of the segment connects smoothly with the beginning, and the DFT behaves well:

```
duration = signal.period * 30
wave = signal.make_wave(duration)
spectrum = wave.make_spectrum()
```

Figure 3-4 (left) shows the result. As expected, there is a single peak at 440 Hz.

But if the duration is not a multiple of the period, bad things happen. With duration = signal.period * 30.25, the signal starts at 0 and ends at 1.

Figure 3-4 (middle) shows the spectrum of this segment. Again, the peak is at 440 Hz, but now there are additional components spread out from 240 to 640 Hz. This spread is called **spectral leakage**, because some of the energy that is actually at the fundamental frequency leaks into other frequencies.

Figure 3-4 (right) shows the spectrum of the windowed signal. Windowing has reduced leakage substantially, but not completely.

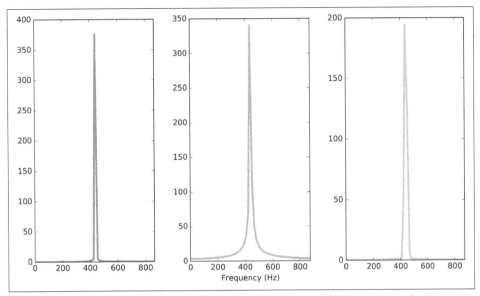

Figure 3-4. Spectrums of a periodic segment of a sinusoid (left), a non-periodic segment (middle), and a windowed non-periodic segment (right).

In this example, leakage happens because we are using the DFT on a segment that becomes discontinuous when we treat it as periodic.

Windowing

We can reduce leakage by smoothing out the discontinuity between the beginning and end of the segment, and one way to do that is **windowing**.

A "window" is a function designed to transform a non-periodic segment into something that can pass for periodic. Figure 3-5 (top) shows a segment where the end does not connect smoothly to the beginning.

Figure 3-5 (middle) shows a "Hamming window", one of the more common window functions. No window function is perfect, but some can be shown to be optimal for different applications, and Hamming is a good, all-purpose window.

Figure 3-5 (bottom) shows the result of multiplying the window by the original signal. Where the window is close to 1, the signal is unchanged. Where the window is close to 0, the signal is attenuated. Because the window tapers at both ends, the end of the segment connects smoothly to the beginning.

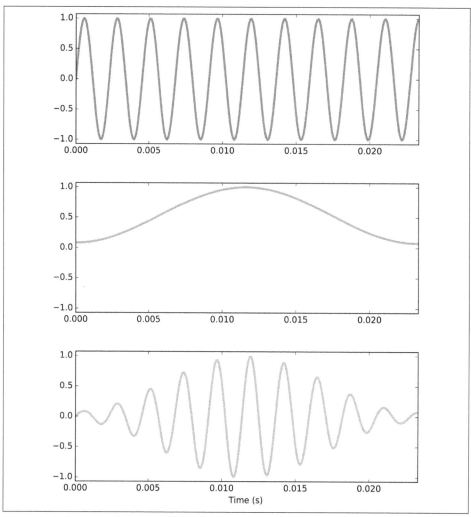

Figure 3-5. Segment of a sinusoid (top), Hamming window (middle), and product of the segment and the window (bottom).

Here's what the code looks like. Wave provides window, which applies a Hamming window:

```
#class Wave:
    def window(self, window):
        self.ys *= window
```

And NumPy provides hamming, which computes a Hamming window with a given length:

```
window = np.hamming(len(wave))
wave.window(window)
```

NumPy provides functions to compute other window functions, including `bartlett`, `blackman`, `hanning`, and `kaiser`. One of the exercises at the end of this chapter asks you to experiment with these other windows.

Implementing Spectrograms

Now that we understand windowing, we can understand the implementation of `make_spectrogram`. Here is the `Wave` method that computes spectrograms:

```
#class Wave:
    def make_spectrogram(self, seg_length):
        window = np.hamming(seg_length)
        i, j = 0, seg_length
        step = seg_length / 2

        spec_map = {}

        while j < len(self.ys):
            segment = self.slice(i, j)
            segment.window(window)

            t = (segment.start + segment.end) / 2
            spec_map[t] = segment.make_spectrum()

            i += step
            j += step

        return Spectrogram(spec_map, seg_length)
```

This is the longest function in the book, so if you can handle this, you can handle anything.

The parameter, `self`, is a `Wave` object. `seg_length` is the number of samples in each segment.

`window` is a Hamming window with the same length as the segments.

`i` and `j` are the slice indices that select segments from the wave. `step` is the offset between segments. Since `step` is half of `seg_length`, the segments overlap by half. Figure 3-6 shows what these overlapping windows look like.

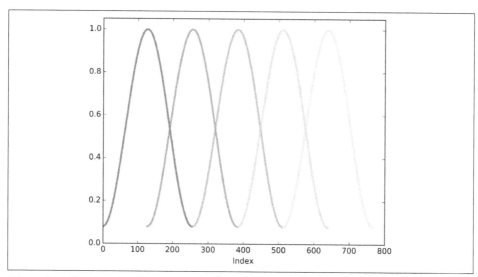

Figure 3-6. Overlapping Hamming windows.

spec_map is a dictionary that maps from a timestamp to a Spectrum.

Inside the while loop, we select a slice from the wave and apply the window; then we construct a Spectrum object and add it to spec_map. The nominal time of each segment, t, is the midpoint.

Then we advance i and j, and continue as long as j doesn't go past the end of the Wave.

Finally, the method constructs and returns a Spectrogram object. Here is the definition of the class:

```
class Spectrogram(object):
    def __init__(self, spec_map, seg_length):
        self.spec_map = spec_map
        self.seg_length = seg_length
```

Like many init methods, this one just stores the parameters as attributes.

Spectrogram provides plot, which generates a pseudocolor plot with time along the x-axis and frequency along the y-axis.

And that's how Spectrograms are implemented.

Exercises

Solutions to these exercises are in chap03soln.ipynb.

Exercise 3-1.

Run and listen to the examples in `chap03.ipynb`, which is in the repository for this book, and also available at *http://tinyurl.com/thinkdsp03*.

In the leakage example, try replacing the Hamming window with one of the other windows provided by NumPy, and see what effect they have on leakage. See *http://docs.scipy.org/doc/numpy/reference/routines.window.html*.

Exercise 3-2.

Write a class called `SawtoothChirp` that extends `Chirp` and overrides `evaluate` to generate a sawtooth waveform with frequency that increases (or decreases) linearly.

Hint: combine the `evaluate` functions from `Chirp` and `SawtoothSignal`.

Draw a sketch of what you think the spectrogram of this signal looks like, and then plot it. The effect of aliasing should be visually apparent, and if you listen carefully, you can hear it.

Exercise 3-3.

Make a sawtooth chirp that sweeps from 2500 to 3000 Hz, then use it to make a wave with duration 1 s and frame rate 20 kHz. Draw a sketch of what you think the spectrum will look like. Then plot the spectrum and see if you got it right.

Exercise 3-4.

In musical terminology, a "glissando" is a note that slides from one pitch to another, so it is similar to a chirp.

Find or make a recording of a glissando and plot a spectrogram of the first few seconds. One suggestion: George Gershwin's *Rhapsody in Blue*, which you can download from *http://archive.org/details/rhapblue11924*, starts with a famous clarinet glissando.

Exercise 3-5.

A trombone player can play a glissando by extending the trombone slide while blowing continuously. As the slide extends, the total length of the tube gets longer, and the resulting pitch is inversely proportional to length.

Assuming that the player moves the slide at a constant speed, how does frequency vary with time?

Write a class called `TromboneGliss` that extends `Chirp` and provides `evaluate`. Make a wave that simulates a trombone glissando from C3 up to F3 and back down to C3. C3 is 262 Hz; F3 is 349 Hz.

Plot a spectrogram of the resulting wave. Is a trombone glissando more like a linear or an exponential chirp?

Exercise 3-6.

Make or find a recording of a series of vowel sounds and look at the spectrogram. Can you identify different vowels?

## CHAPTER 4

Noise

In English, "noise" means an unwanted or unpleasant sound. In the context of signal processing, it has two different senses:

1. As in English, it can mean an unwanted signal of any kind. If two signals interfere with each other, each signal would consider the other to be noise.

2. "Noise" also refers to a signal that contains components at many frequencies, so it lacks the harmonic structure of the periodic signals we saw in previous chapters.

This chapter is about the second kind.

The code for this chapter is in chap04.ipynb, which is in the repository for this book (see "Using the Code" on page viii). You can also view it at *http://tinyurl.com/thinkdsp04.*

Uncorrelated Noise

The simplest way to understand noise is to generate it, and the simplest kind to generate is **uncorrelated uniform noise** (UU noise). "Uniform" means the signal contains random values from a uniform distribution; that is, every value in the range is equally likely. "Uncorrelated" means that the values are independent; that is, knowing one value provides no information about the others.

Here's a class that represents UU noise:

```
class UncorrelatedUniformNoise(_Noise):

    def evaluate(self, ts):
        ys = np.random.uniform(-self.amp, self.amp, len(ts))
        return ys
```

`UncorrelatedUniformNoise` inherits from `_Noise`, which inherits from `Signal`.

As usual, the evaluate function takes `ts`, the times when the signal should be evaluated. It uses `np.random.uniform`, which generates values from a uniform distribution. In this example, the values are in the range between `-amp` and `amp`.

The following example generates UU noise with duration 0.5 seconds at 11,025 samples per second:

```
signal = thinkdsp.UncorrelatedUniformNoise()
wave = signal.make_wave(duration=0.5, framerate=11025)
```

If you play this wave, it sounds like the static you hear if you tune a radio between channels. Figure 4-1 shows what the waveform looks like. As expected, it looks pretty random.

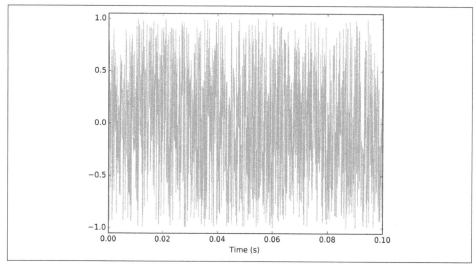

Figure 4-1. Waveform of uncorrelated uniform noise.

Now let's take a look at the spectrum:

```
spectrum = wave.make_spectrum()
spectrum.plot_power()
```

`Spectrum.plot_power` is similar to `Spectrum.plot`, except that it plots power instead of amplitude. Power is the square of amplitude. I am switching from amplitude to power in this chapter because it is more conventional in the context of noise.

Figure 4-2 shows the result. Like the signal, the spectrum looks pretty random. In fact, it *is* random, but we have to be more precise about the word "random". There are at least three things we might like to know about a noise signal or its spectrum:

Distribution

> The distribution of a random signal is the set of possible values and their probabilities. For example, in the uniform noise signal, the set of values is the range from –1 to 1, and all values have the same probability. An alternative is **Gaussian noise**, where the set of values is the range from negative to positive infinity, but values near 0 are the most likely, with probability that drops off according to the Gaussian or "bell" curve.

Correlation

> Is each value in the signal independent from the others, or are there dependencies between them? In UU noise, the values are independent. An alternative is **Brownian noise**, where each value is the sum of the previous value and a random "step". So if the value of the signal is high at a particular point in time, we expect it to stay high, and if it is low, we expect it to stay low.

Relationship between power and frequency

> In the spectrum of UU noise, the power at all frequencies is drawn from the same distribution; that is, the average power is the same for all frequencies. An alternative is **pink noise**, where power is inversely related to frequency; that is, the power at frequency f is drawn from a distribution whose mean is proportional to $1/f$.

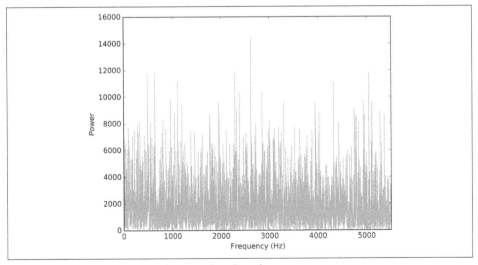

Figure 4-2. Power spectrum of uncorrelated uniform noise.

Integrated Spectrum

For UU noise we can see the relationship between power and frequency more clearly by looking at the **integrated spectrum**, which is a function of frequency, f, that shows the cumulative power in the spectrum up to f.

Spectrum provides a method that computes the IntegratedSpectrum:

```
def make_integrated_spectrum(self):
    cs = np.cumsum(self.power)
    cs /= cs[-1]
    return IntegratedSpectrum(cs, self.fs)
```

self.power is a NumPy array containing the power for each frequency. np.cumsum computes the cumulative sum of the powers. Dividing through by the last element normalizes the integrated spectrum so it runs from 0 to 1.

The result is an IntegratedSpectrum. Here is the class definition:

```
class IntegratedSpectrum(object):
    def __init__(self, cs, fs):
        self.cs = cs
        self.fs = fs
```

Like Spectrum, IntegratedSpectrum provides plot_power, so we can compute and plot the integrated spectrum like this:

```
integ = spectrum.make_integrated_spectrum()
integ.plot_power()
```

The result, shown in Figure 4-3, is a straight line, which indicates that power at all frequencies is constant, on average. Noise with equal power at all frequencies is called **white noise** by analogy with light, because an equal mixture of light at all visible frequencies is white.

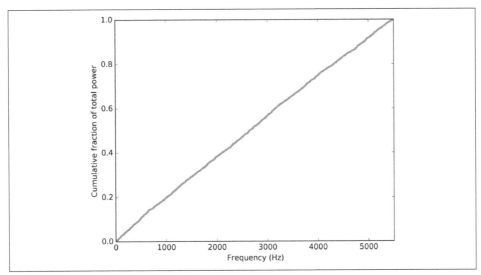

Figure 4-3. Integrated spectrum of uncorrelated uniform noise.

Brownian Noise

UU noise is uncorrelated, which means that each value does not depend on the others. An alternative is Brownian noise, in which each value is the sum of the previous value and a random "step".

It is called "Brownian" by analogy with Brownian motion, in which a particle suspended in a fluid moves apparently at random, due to unseen interactions with the fluid. Brownian motion is often described using a **random walk**, which is a mathematical model of a path where the distance between steps is characterized by a random distribution.

In a one-dimensional random walk, the particle moves up or down by a random amount at each timestep. The location of the particle at any point in time is the sum of all previous steps.

This observation suggests a way to generate Brownian noise: generate uncorrelated random steps and then add them up. Here is a class definition that implements this algorithm:

```
class BrownianNoise(_Noise):

    def evaluate(self, ts):
        dys = np.random.uniform(-1, 1, len(ts))
        ys = np.cumsum(dys)
        ys = normalize(unbias(ys), self.amp)
        return ys
```

`evaluate` uses `np.random.uniform` to generate an uncorrelated signal and `np.cumsum` to compute their cumulative sum.

Since the sum is likely to escape the range from −1 to 1, we have to use `unbias` to shift the mean to 0, and `normalize` to get the desired maximum amplitude.

Here's the code that generates a `BrownianNoise` object and plots the waveform:

```
signal = thinkdsp.BrownianNoise()
wave = signal.make_wave(duration=0.5, framerate=11025)
wave.plot()
```

Figure 4-4 shows the result. The waveform wanders up and down, but there is a clear correlation between successive values. When the amplitude is high, it tends to stay high, and vice versa.

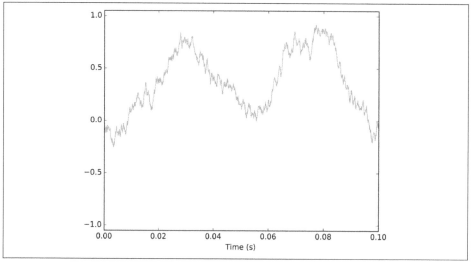

Figure 4-4. Waveform of Brownian noise.

If you plot the spectrum of Brownian noise on a linear scale, as in Figure 4-5 (left), it doesn't look like much. Nearly all of the power is at the lowest frequencies; the higher frequency components are not visible.

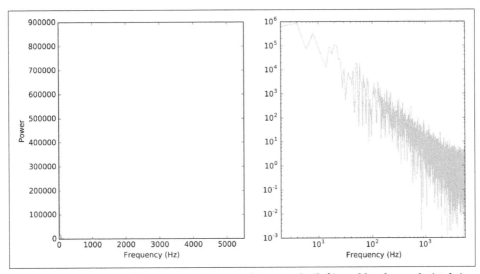

Figure 4-5. Spectrum of Brownian noise on a linear scale (left) and log-log scale (right).

To see the shape of the spectrum more clearly, we can plot power and frequency on a log-log scale. Here's the code:

```
spectrum = wave.make_spectrum()
spectrum.plot_power(linewidth=1, alpha=0.5)
thinkplot.config(xscale='log', yscale='log')
```

The result is in Figure 4-5 (right). The relationship between power and frequency is noisy, but roughly linear.

Spectrum provides estimate_slope, which uses SciPy to compute a least squares fit to the power spectrum:

```
#class Spectrum

    def estimate_slope(self):
        x = np.log(self.fs[1:])
        y = np.log(self.power[1:])
        t = scipy.stats.linregress(x,y)
        return t
```

It discards the first component of the spectrum because this component corresponds to $f = 0$, and log 0 is undefined.

estimate_slope returns the result from scipy.stats.linregress, which is an object that contains the estimated slope and intercept, coefficient of determination (R^2), p-value, and standard error. For our purposes, we only need the slope.

For Brownian noise, the slope of the power spectrum is –2 (we'll see why in Chapter 9), so we can write this relationship:

$$\log P = k - 2 \log f$$

where P is power, f is frequency, and k is the intercept of the line, which is not important for our purposes. Exponentiating both sides yields:

$$P = K / f^2$$

where K is e^k, but still not important. More relevant is that power is proportional to $1/f^2$, which is characteristic of Brownian noise.

Brownian noise is also called **red noise**, for the same reason that white noise is called "white". If you combine visible light with power proportional to $1/f^2$, most of the power will be at the low-frequency end of the spectrum, which is red. Brownian noise is also sometimes called "brown noise", but I think that's confusing, so I won't use it.

Pink Noise

For red noise, the relationship between frequency and power is:

$$P = K / f^2$$

There is nothing special about the exponent 2. More generally, we can synthesize noise with any exponent, β:

$$P = K / f^\beta$$

When $\beta = 0$, power is constant at all frequencies, so the result is white noise. When $\beta = 2$ the result is red noise.

When β is between 0 and 2, the result is between white and red noise, so it is called **pink noise**.

There are several ways to generate pink noise. The simplest is to generate white noise and then apply a low-pass filter with the desired exponent. thinkdsp provides a class that represents a pink noise signal:

```
class PinkNoise(_Noise):

    def __init__(self, amp=1.0, beta=1.0):
        self.amp = amp
        self.beta = beta
```

amp is the desired amplitude of the signal. beta is the desired exponent. PinkNoise provides make_wave, which generates a Wave:

```
def make_wave(self, duration=1, start=0, framerate=11025):
    signal = UncorrelatedUniformNoise()
    wave = signal.make_wave(duration, start, framerate)
    spectrum = wave.make_spectrum()

    spectrum.pink_filter(beta=self.beta)

    wave2 = spectrum.make_wave()
    wave2.unbias()
    wave2.normalize(self.amp)
    return wave2
```

duration is the length of the wave in seconds. start is the start time of the wave; it is included so that make_wave has the same interface for all types of signal, but for random noise, start time is irrelevant. framerate is the number of samples per second.

make_wave creates a white noise wave, computes its spectrum, applies a filter with the desired exponent, and then converts the filtered spectrum back to a wave. Then it unbiases and normalizes the wave.

Spectrum provides pink_filter:

```
def pink_filter(self, beta=1.0):
    denom = self.fs ** (beta/2.0)
    denom[0] = 1
    self.hs /= denom
```

pink_filter divides each element of the spectrum by $f^{\beta/2}$. Since power is the square of amplitude, this operation divides the power at each component by f^{β}. It treats the component at $f = 0$ as a special case, partly to avoid dividing by 0 and partly because this element represents the bias of the signal, which we are going to set to 0 anyway.

Figure 4-6 shows the resulting waveform. Like that of Brownian noise, it wanders up and down in a way that suggests correlation between successive values, but at least visually, it looks more random. In the next chapter we will come back to this observation and I will be more precise about what I mean by "correlation" and "more random".

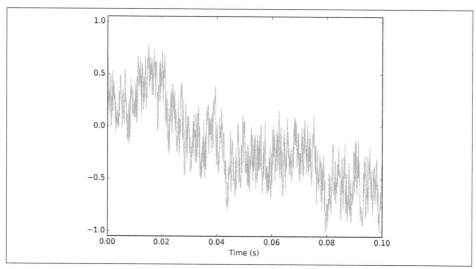

Figure 4-6. Waveform of pink noise with β = 1.

Finally, Figure 4-7 shows a spectrum for white, pink, and red noise on the same log-log scale. The relationship between the exponent, *β*, and the slope of the spectrum is apparent in this figure.

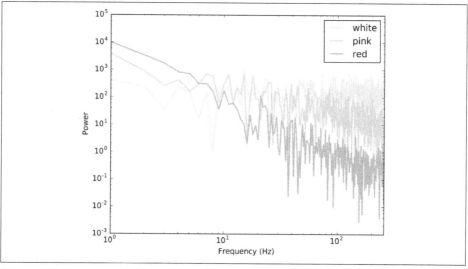

Figure 4-7. Spectrum of white, pink, and red noise on a log-log scale.

Gaussian Noise

We started with uncorrelated uniform (UU) noise and showed that, because its spectrum has equal power at all frequencies, on average, UU noise is white.

But when people talk about "white noise", they don't always mean UU noise. In fact, more often they mean uncorrelated Gaussian (UG) noise.

thinkdsp provides an implementation of UG noise:

```
class UncorrelatedGaussianNoise(_Noise):

    def evaluate(self, ts):
        ys = np.random.normal(0, self.amp, len(ts))
        return ys
```

np.random.normal returns a NumPy array of values from a Gaussian distribution, in this case with mean 0 and standard deviation self.amp. In theory the range of values is from negative to positive infinity, but we expect about 99% of the values to be between −3 and 3.

UG noise is similar in many ways to UU noise. The spectrum has equal power at all frequencies, on average, so UG is also white. And it has one other interesting property: the spectrum of UG noise is also UG noise. More precisely, the real and imaginary parts of the spectrum are uncorrelated Gaussian values.

To test that claim, we can generate the spectrum of UG noise and then generate a "normal probability plot", which is a graphical way to test whether a distribution is Gaussian:

```
signal = thinkdsp.UncorrelatedGaussianNoise()
wave = signal.make_wave(duration=0.5, framerate=11025)
spectrum = wave.make_spectrum()

thinkstats2.NormalProbabilityPlot(spectrum.real)
thinkstats2.NormalProbabilityPlot(spectrum.imag)
```

NormalProbabilityPlot is provided by thinkstats2, which is included in the repository for this book. If you are not familiar with normal probability plots, you can read about them in Chapter 5 of *Think Stats* at *http://thinkstats2.com*.

Figure 4-8 shows the results. The gray lines show a linear model fit to the data; the dark lines show the data.

A straight line on a normal probability plot indicates that the data come from a Gaussian distribution. Except for some random variation at the extremes, these lines are straight, which indicates that the spectrum of UG noise is UG noise.

The spectrum of UU noise is also UG noise, at least approximately. In fact, by the Central Limit Theorem, the spectrum of almost any uncorrelated noise is approximately Gaussian, as long as the distribution has finite mean and standard deviation, and the number of samples is large.

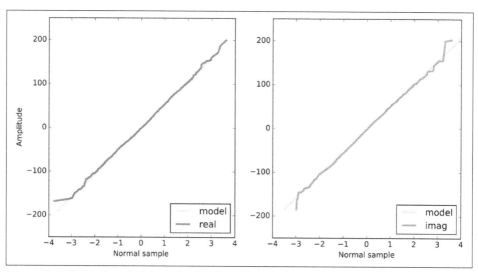

Figure 4-8. Normal probability plot for the real and imaginary parts of the spectrum of Gaussian noise.

Exercises

Solutions to these exercises are in `chap04soln.ipynb`.

Exercise 4-1.

A Soft Murmur is a website that plays a mixture of natural noise sources, including rain, waves, wind, etc. At *http://asoftmurmur.com/about/* you can find their list of recordings, most of which are at *http://freesound.org*.

Download a few of these files and compute the spectrum of each signal. Does the power spectrum look like white noise, pink noise, or Brownian noise? How does the spectrum vary over time?

Exercise 4-2.

In a noise signal, the mixture of frequencies changes over time. In the long run, we expect the power at all frequencies to be equal, but in any sample, the power at each frequency is random.

To estimate the long-term average power at each frequency, we can break a long signal into segments, compute the power spectrum for each segment, and then compute the average across the segments. You can read more about this algorithm at *http://en.wikipedia.org/wiki/Bartlett's_method*.

Implement Bartlett's method and use it to estimate the power spectrum for a noise wave. Hint: look at the implementation of make_spectrogram.

Exercise 4-3.

At *http://www.coindesk.com/price*, you can download historical data on the daily price of a BitCoin as a CSV file. Read this file and compute the spectrum of BitCoin prices as a function of time. Does it resemble white, pink, or Brownian noise?

Exercise 4-4.

A Geiger counter is a device that detects radiation. When an ionizing particle strikes the detector, it outputs a surge of current. The total output at a point in time can be modeled as uncorrelated Poisson (UP) noise, where each sample is a random quantity from a Poisson distribution, which corresponds to the number of particles detected during an interval.

Write a class called UncorrelatedPoissonNoise that inherits from thinkdsp._Noise and provides evaluate. It should use np.random.poisson to generate random values from a Poisson distribution. The parameter of this function, lam, is the average number of particles during each interval. You can use the attribute amp to specify lam. For example, if the frame rate is 10 kHz and amp is 0.001, we expect about 10 "clicks" per second.

Generate about a second of UP noise and listen to it. For low values of amp, like 0.001, it should sound like a Geiger counter. For higher values it should sound like white noise. Compute and plot the power spectrum to see whether it looks like white noise.

Exercise 4-5.

The algorithm in this chapter for generating pink noise is conceptually simple but computationally expensive. There are more efficient alternatives, like the Voss–McCartney algorithm. Research this method, implement it, compute the spectrum of the result, and confirm that it has the desired relationship between power and frequency.

## CHAPTER 5

Autocorrelation

In the previous chapter I characterized white noise as "uncorrelated", which means that each value is independent of the others, and Brownian noise as "correlated", because each value depends on the preceding value. In this chapter I define these terms more precisely and present the **autocorrelation function**, which is a useful tool for signal analysis.

The code for this chapter is in `chap05.ipynb`, which is in the repository for this book (see "Using the Code" on page viii). You can also view it at *http://tinyurl.com/thinkdsp05*.

Correlation

In general, correlation between variables means that if you know the value of one, you have some information about the other. There are several ways to quantify correlation, but the most common is the Pearson product-moment correlation coefficient, usually denoted ρ. For two variables, x and y, that each contain N values:

$$\rho = \frac{\Sigma_i \left(x_i - \mu_x\right)\left(y_i - \mu_y\right)}{N\sigma_x\sigma_y}$$

where μ_x and μ_y are the means of x and y, and σ_x and σ_y are their standard deviations.

Pearson's correlation is always between -1 and $+1$ (including both). If ρ is positive, we say that the correlation is positive, which means that when one variable is high, the other tends to be high. If ρ is negative, the correlation is negative, so when one variable is high, the other tends to be low.

The magnitude of ρ indicates the strength of the correlation. If ρ is 1 or -1, the variables are perfectly correlated, which means that if you know one, you can make a perfect prediction about the other. If ρ is near zero, the correlation is probably weak, so if you know one, it doesn't tell you much about the others.

I say "probably weak" because it is also possible that there is a nonlinear relationship that is not captured by the coefficient of correlation. Nonlinear relationships are often important in statistics, but less often relevant for signal processing, so I won't say more about them here.

Python provides several ways to compute correlations. np.corrcoef takes any number of variables and computes a **correlation matrix** that includes correlations between each pair of variables.

I'll present an example with only two variables. First, I define a function that constructs sine waves with different phase offsets:

```
def make_sine(offset):
    signal = thinkdsp.SinSignal(freq=440, offset=offset)
    wave = signal.make_wave(duration=0.5, framerate=10000)
    return wave
```

Next I instantiate two waves with different offsets:

```
wave1 = make_sine(offset=0)
wave2 = make_sine(offset=1)
```

Figure 5-1 shows what the first few periods of these waves look like. When one wave is high, the other is usually high, so we expect them to be correlated:

```
>>> corr_matrix = np.corrcoef(wave1.ys, wave2.ys, ddof=0)
[[ 1.    0.54]
 [ 0.54  1.  ]]
```

The option ddof=0 indicates that corrcoef should divide by N, as in the equation above, rather than use the default, $N - 1$.

The result is a correlation matrix. The first element is the correlation of wave1 with itself, which is always 1. Similarly, the last element is the correlation of wave2 with itself.

The off-diagonal elements contain the value we're interested in, the correlation of wave1 and wave2. The value 0.54 indicates that the strength of the correlation is moderate.

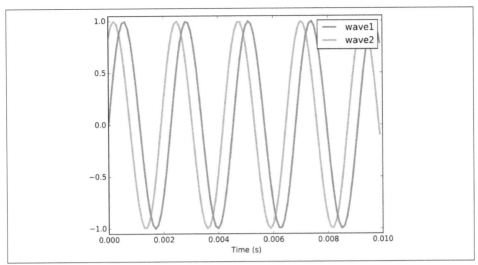

Figure 5-1. Two sine waves that differ by a phase offset of 1 radian; their coefficient of correlation is 0.54.

As the phase offset increases, this correlation decreases until the waves are 180 degrees out of phase, which yields correlation –1. Then it increases until the offset differs by 360 degrees. At that point we have come full circle and the correlation is 1.

Figure 5-2 shows the relationship between correlation and phase offset for a sine wave. The shape of that curve should look familiar; it is a cosine.

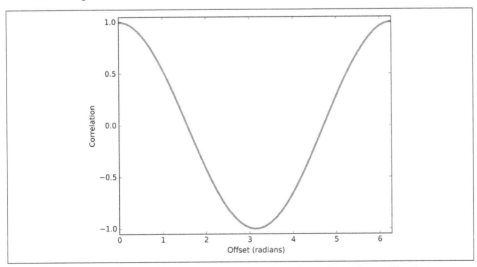

Figure 5-2. The correlation of two sine waves as a function of the phase offset between them. The result is a cosine.

thinkdsp provides a simple interface for computing the correlation between waves:

```
>>> wave1.corr(wave2)
0.54
```

Serial Correlation

Signals often represent measurements of quantities that vary in time. For example, the sound signals we've worked with represent measurements of voltage (or current), which correspond to the changes in air pressure we perceive as sound.

Measurements like these almost always have serial correlation, which is the correlation between each element and the next (or the previous). To compute serial correlation, we can shift a signal and then compute the correlation of the shifted version with the original:

```
def serial_corr(wave, lag=1):
    n = len(wave)
    y1 = wave.ys[lag:]
    y2 = wave.ys[:n-lag]
    corr = np.corrcoef(y1, y2, ddof=0)[0, 1]
    return corr
```

serial_corr takes a Wave object and lag, which is the integer number of places to shift the wave. It computes the correlation of the wave with a shifted version of itself.

We can test this function with the noise signals from the previous chapter. We expect UU noise to be uncorrelated, based on the way it's generated (not to mention the name):

```
signal = thinkdsp.UncorrelatedGaussianNoise()
wave = signal.make_wave(duration=0.5, framerate=11025)
serial_corr(wave)
```

When I ran this example, I got 0.006, which indicates a very small serial correlation. You might get a different value when you run it, but it should be comparably small.

In a Brownian noise signal, each value is the sum of the previous value and a random "step", so we expect a strong serial correlation:

```
signal = thinkdsp.BrownianNoise()
wave = signal.make_wave(duration=0.5, framerate=11025)
serial_corr(wave)
```

Sure enough, the result I got is greater than 0.999.

Since pink noise is in some sense between Brownian noise and UU noise, we might expect an intermediate correlation:

```
signal = thinkdsp.PinkNoise(beta=1)
wave = signal.make_wave(duration=0.5, framerate=11025)
serial_corr(wave)
```

With parameter $\beta = 1$, I got a serial correlation of 0.851. As we vary the parameter from $\beta = 0$, which is uncorrelated noise, to $\beta = 2$, which is Brownian, serial correlation ranges from 0 to almost 1, as shown in Figure 5-3.

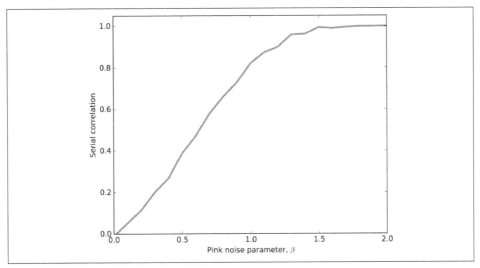

Figure 5-3. Serial correlation for pink noise with a range of parameters.

Autocorrelation

In the previous section we computed the correlation between each value and the next, so we shifted the elements of the array by 1. But we can easily compute serial correlations with different lags.

You can think of serial_corr as a function that maps from each value of lag to the corresponding correlation, and we can evaluate that function by looping through values of lag:

```
def autocorr(wave):
    lags = range(len(wave.ys)//2)
    corrs = [serial_corr(wave, lag) for lag in lags]
    return lags, corrs
```

autocorr takes a Wave object and returns the autocorrelation function as a pair of sequences: lags is a sequence of integers from 0 to half the length of the wave; corrs is the sequence of serial correlations for each lag.

Figure 5-4 shows autocorrelation functions for pink noise with three values of β. For low values of β, the signal is less correlated, and the autocorrelation function drops off to zero quickly. For larger values, serial correlation is stronger and drops off more slowly. With $\beta = 1.7$ serial correlation is strong even for long lags; this phenomenon is called **long-range dependence**, because it indicates that each value in the signal depends on many preceding values.

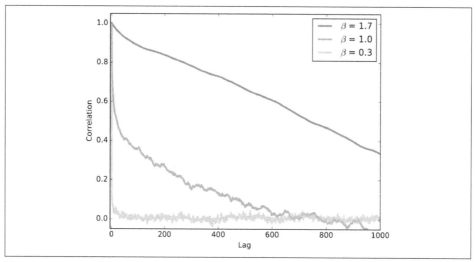

Figure 5-4. Autocorrelation functions for pink noise with a range of parameters.

Autocorrelation of Periodic Signals

The autocorrelation of pink noise has interesting mathematical properties, but limited applications. The autocorrelation of periodic signals is more useful.

As an example, I downloaded from *https://freesound.org* a recording of someone singing a chirp; the repository for this book includes the file (*https://github.com/Allen Downey/ThinkDSP/blob/master/code/28042__bcjordan__voicedownbew.wav*). You can use the Jupyter notebook for this chapter, chap05.ipynb, to play it.

Figure 5-5 shows the spectrogram of this wave. The fundamental frequency and some of the harmonics show up clearly. The chirp starts near 500 Hz and drops down to about 300 Hz, roughly from C5 to E4.

To estimate pitch at a particular point in time, we could use the spectrum, but it doesn't work very well. To see why not, I'll take a short segment from the wave and plot its spectrum:

```
duration = 0.01
segment = wave.segment(start=0.2, duration=duration)
spectrum = segment.make_spectrum()
spectrum.plot(high=1000)
```

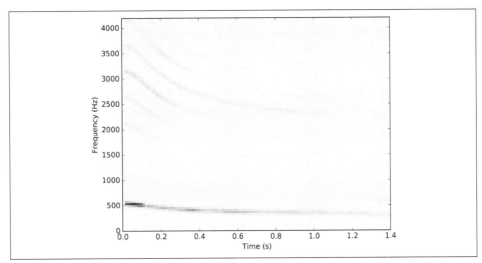

Figure 5-5. Spectrogram of a vocal chirp.

This segment starts at 0.2 seconds and lasts 0.01 seconds. Figure 5-6 shows its spectrum. There is a clear peak near 400 Hz, but it is hard to identify the pitch precisely. The length of the segment is 441 samples at a frame rate of 44,100 Hz, so the frequency resolution is 100 Hz (see "The Gabor Limit" on page 31). That means the estimated pitch might be off by 50 Hz; in musical terms, the range from 350 Hz to 450 Hz is about 5 semitones, which is a big difference!

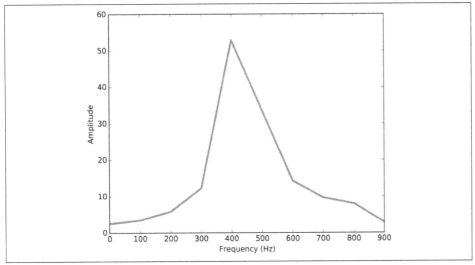

Figure 5-6. Spectrum of a segment from a vocal chirp.

We could get better frequency resolution by taking a longer segment, but since the pitch is changing over time, we would also get "motion blur"; that is, the peak would spread between the start and end pitch of the segment, as we saw in "Spectrum of a Chirp" on page 29.

We can estimate pitch more precisely using autocorrelation. If a signal is periodic, we expect the autocorrelation to spike when the lag equals the period.

To show why that works, I'll plot two segments from the same recording:

```
def plot_shifted(wave, offset=0.001, start=0.2):
    thinkplot.preplot(2)
    segment1 = wave.segment(start=start, duration=0.01)
    segment1.plot(linewidth=2, alpha=0.8)

    segment2 = wave.segment(start=start-offset, duration=0.01)
    segment2.shift(offset)
    segment2.plot(linewidth=2, alpha=0.4)

    corr = segment1.corr(segment2)
    text = r'$\rho =$ %.2g' % corr
    thinkplot.text(segment1.start+0.0005, -0.8, text)
    thinkplot.config(xlabel='Time (s)')
```

One segment starts at 0.2 seconds; the other starts 0.0023 seconds later. Figure 5-7 shows the result. The segments are similar, and their correlation is 0.99. This result suggests that the period is near 0.0023 seconds, which corresponds to a frequency of 435 Hz.

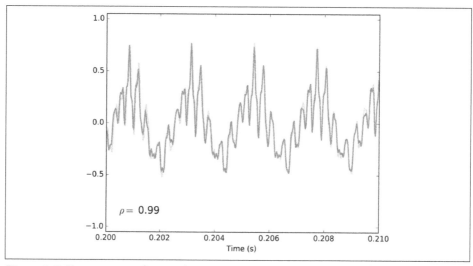

Figure 5-7. Two segments from a chirp, one starting 0.0023 seconds after the other.

For this example, I estimated the period by trial and error. To automate the process, we can use the autocorrelation function:

```
lags, corrs = autocorr(segment)
thinkplot.plot(lags, corrs)
```

Figure 5-8 shows the autocorrelation function for the segment starting at $t = 0.2$ seconds. The first peak occurs at `lag=101`. We can compute the frequency that corresponds to that period like this:

```
period = lag / segment.framerate
frequency = 1 / period
```

The estimated fundamental frequency is 437 Hz. To evaluate the precision of the estimate, we can run the same computation with lags 100 and 102, which correspond to frequencies 432 and 441 Hz. The frequency precision using autocorrelation is less than 10 Hz, compared with 100 Hz using the spectrum. In musical terms, the expected error is about 30 cents (a third of a semitone).

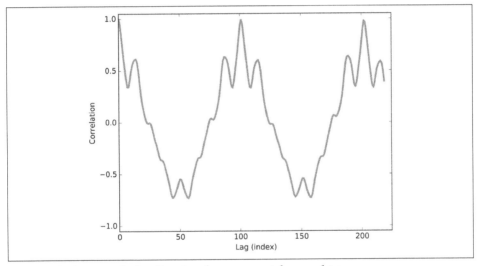

Figure 5-8. Autocorrelation function for a segment from a chirp.

Correlation as Dot Product

I started the chapter with this definition of Pearson's correlation coefficient:

$$\rho = \frac{\Sigma_i (x_i - \mu_x)(y_i - \mu_y)}{N \sigma_x \sigma_y}$$

Then I used ρ to define serial correlation and autocorrelation. That's consistent with how these terms are used in statistics, but in the context of signal processing, the definitions are a little different.

In signal processing, we are often working with unbiased signals, where the mean is 0, and normalized signals, where the standard deviation is 1. In that case, the definition of ρ simplifies to:

$$\rho = \frac{1}{N} \sum_i x_i y_i$$

And it is common to simplify even further:

$$r = \sum_i x_i y_i$$

This definition of correlation is not "standardized", so it doesn't generally fall between -1 and 1. But it has other useful properties.

If you think of x and y as vectors, you might recognize this formula as the **dot product**, $x \cdot y$. See *http://en.wikipedia.org/wiki/Dot_product*.

The dot product indicates the degree to which the signals are similar. If they are normalized so their standard deviations are 1:

$$x \cdot y = \cos \theta$$

where θ is the angle between the vectors. And that explains why Figure 5-2 is a cosine curve.

Using NumPy

NumPy provides a function, `correlate`, that computes the correlation of two functions or the autocorrelation of one function. We can use it to compute the autocorrelation of the segment from the previous section:

```
corrs2 = np.correlate(segment.ys, segment.ys, mode='same')
```

The option `mode` tells `correlate` what range of `lag` to use. With the value `'same'`, the range is from $-N/2$ to $N/2$, where N is the length of the wave array.

Figure 5-9 shows the result. It is symmetric because the two signals are identical, so a negative lag on one has the same effect as a positive lag on the other. To compare with the results from `autocorr`, we can select the second half:

```
N = len(corrs2)
half = corrs2[N//2:]
```

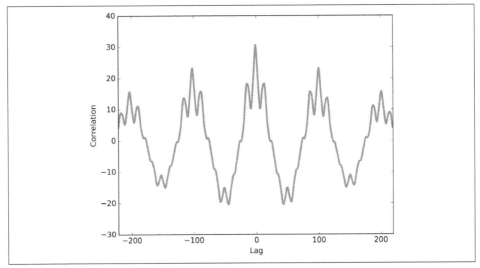

Figure 5-9. Autocorrelation function computed with np.correlate.

If you compare Figure 5-9 to Figure 5-8, you'll notice that the correlations computed by np.correlate get smaller as the lags increase. That's because np.correlate uses the unstandardized definition of correlation; as the lag gets bigger, the number of points in the overlap between the two signals gets smaller, so the magnitude of the correlations decreases.

We can correct that by dividing through by the lengths:

```
lengths = range(N, N//2, -1)
half /= lengths
```

Finally, we can normalize the results so the correlation with lag=0 is 1:

```
half /= half[0]
```

With these adjustments, the results computed by autocorr and np.correlate are nearly the same. They still differ by 1–2%. The reason is not important, but if you are curious: autocorr standardizes the correlations independently for each lag; for np.correlate, we standardized them all at the end.

More importantly, now you know what autocorrelation is, how to use it to estimate the fundamental period of a signal, and two ways to compute it.

Exercises

Solutions to these exercises are in `chap05soln.ipynb`.

Exercise 5-1.

The Jupyter notebook for this chapter, `chap05.ipynb`, includes an interaction that lets you compute autocorrelations for different lags. Use this interaction to estimate the pitch of the vocal chirp for a few different start times.

Exercise 5-2.

The example code in `chap05.ipynb` shows how to use autocorrelation to estimate the fundamental frequency of a periodic signal. Encapsulate this code in a function called `estimate_fundamental`, and use it to track the pitch of a recorded sound.

To see how well it works, try superimposing your pitch estimates on a spectrogram of the recording.

Exercise 5-3.

If you did the exercises in the previous chapter, you downloaded the historical prices of BitCoins and estimated the power spectrum of the price changes. Using the same data, compute the autocorrelation of BitCoin prices. Does the autocorrelation function drop off quickly? Is there evidence of periodic behavior?

Exercise 5-4.

In the repository for this book you will find a Jupyter notebook called `saxo phone.ipynb` that explores autocorrelation, pitch perception, and a phenomenon called the **missing fundamental**. Read through this notebook and run the examples. Try selecting a different segment of the recording and running the examples again.

Vi Hart has an excellent video called "What is up with Noises? (The Science and Mathematics of Sound, Frequency, and Pitch)"; it demonstrates the missing fundamental phenomenon and explains how pitch perception works (at least, to the degree that we know). Watch it at *https://www.youtube.com/watch?v=i_0DXxNeaQ0*.

## CHAPTER 6

Discrete Cosine Transform

The topic of this chapter is the **Discrete Cosine Transform** (DCT), which is used in MP3 and related formats for compressing music; JPEG and similar formats for images; and the MPEG family of formats for video.

The DCT is similar in many ways to the Discrete Fourier Transform (DFT), which we have been using for spectral analysis. Once we learn how the DCT works, it will be easier to explain the DFT.

Here are the steps to get there:

1. We'll start with the synthesis problem: given a set of frequency components and their amplitudes, how can we construct a wave?

2. Next we'll rewrite the synthesis problem using NumPy arrays. This move is good for performance, and also provides insight for the next step.

3. We'll look at the analysis problem: given a signal and a set of frequencies, how can we find the amplitude of each frequency component? We'll start with a solution that is conceptually simple but slow.

4. Finally, we'll use some principles from linear algebra to find a more efficient algorithm. If you already know linear algebra, that's great, but I'll explain what you need as we go.

The code for this chapter is in chap06.ipynb, which is in the repository for this book (see "Using the Code" on page viii). You can also view it at *http://tinyurl.com/thinkdsp06*.

Synthesis

Suppose I give you a list of amplitudes and a list of frequencies, and ask you to construct a signal that is the sum of these frequency components. Using objects in the `thinkdsp` module, there is a simple way to perform this operation, which is called **synthesis**:

```
def synthesize1(amps, fs, ts):
    components = [thinkdsp.CosSignal(freq, amp)
                  for amp, freq in zip(amps, fs)]
    signal = thinkdsp.SumSignal(*components)

    ys = signal.evaluate(ts)
    return ys
```

`amps` is a list of amplitudes, `fs` is the list of frequencies, and `ts` is the sequence of times where the signal should be evaluated.

`components` is a list of `CosSignal` objects, one for each amplitude–frequency pair. `SumSignal` represents the sum of these frequency components.

Finally, `evaluate` computes the value of the signal at each time in `ts`.

We can test this function like this:

```
amps = np.array([0.6, 0.25, 0.1, 0.05])
fs = [100, 200, 300, 400]
framerate = 11025

ts = np.linspace(0, 1, framerate)
ys = synthesize1(amps, fs, ts)
wave = thinkdsp.Wave(ys, framerate)
```

This example makes a signal that contains a fundamental frequency at 100 Hz and three harmonics (100 Hz is a sharp G2). It renders the signal for 1 second at 11,025 frames per second and puts the results into a `Wave` object.

Conceptually, synthesis is pretty simple. But in this form it doesn't help much with **analysis**, which is the inverse problem: given the wave, how do we identify the frequency components and their amplitudes?

Synthesis with Arrays

Here's another way to write `synthesize`:

```
def synthesize2(amps, fs, ts):
    args = np.outer(ts, fs)
    M = np.cos(PI2 * args)
    ys = np.dot(M, amps)
    return ys
```

This function looks very different, but it does the same thing. Let's see how it works:

1. `np.outer` computes the **outer product** of `ts` and `fs`. The result is an array with one row for each element of `ts` and one column for each element of `fs`. Each element in the array is the product of a frequency and a time, ft.

2. We multiply `args` by 2π and apply `cos`, so each element of the result is $\cos(2\pi ft)$. Since the `ts` runs down the columns, each column contains a cosine signal at a particular frequency, evaluated at a sequence of times.

3. `np.dot` multiplies each row of M by `amps`, elementwise, and then adds up the products. In terms of linear algebra, we are multiplying a matrix, M, by a vector, `amps`. In terms of signals, we are computing the weighted sum of frequency components.

Figure 6-1 shows the structure of this computation. Each row of the matrix, M, corresponds to a time from 0.0 to 1.0 seconds; t_n is the time of the nth row. Each column corresponds to a frequency from 100 to 400 Hz; f_k is the frequency of the kth column.

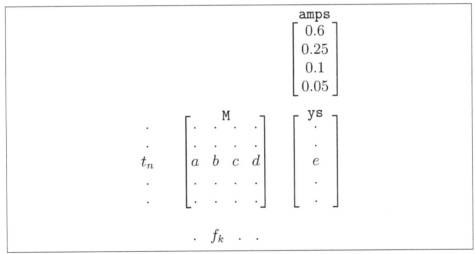

Figure 6-1. Synthesis with arrays.

I labeled the nth row with the letters a through d; as an example, the value of a is $\cos\left[2\pi(100)t_n\right]$.

The result of the dot product, `ys`, is a vector with one element for each row of M. The nth element, labeled e, is the sum of products:

$$e = 0.6a + 0.25b + 0.1c + 0.05d$$

And likewise with the other elements of ys. So each element of ys is the sum of four frequency components, evaluated at a point in time and multiplied by the corresponding amplitudes. And that's exactly what we wanted.

We can use the code from the previous section to check that the two versions of synthesize produce the same results:

```
ys1 = synthesize1(amps, fs, ts)
ys2 = synthesize2(amps, fs, ts)
max(abs(ys1 - ys2))
```

The biggest difference between ys1 and ys2 is about 1e-13, which is what we expect due to floating-point errors.

Writing this computation in terms of linear algebra makes the code smaller and faster. Linear algebra provides concise notation for operations on matrices and vectors. For example, we could write synthesize like this:

$$M = \cos{(2\pi t \otimes f)}$$
$$y = Ma$$

where a is a vector of amplitudes, t is a vector of times, f is a vector of frequencies, and $\otimes$ is the symbol for the outer product of two vectors.

Analysis

Now we are ready to solve the analysis problem. Suppose I give you a wave and tell you that it is the sum of cosines with a given set of frequencies. How would you find the amplitude for each frequency component? In other words, given ys, ts, and fs, can you recover amps?

In terms of linear algebra, the first step is the same as for synthesis: we compute $M = \cos{(2\pi t \otimes f)}$. Then we want to find a so that $y = Ma$; in other words, we want to solve a linear system. NumPy provides linalg.solve, which does exactly that.

Here's what the code looks like:

```
def analyze1(ys, fs, ts):
    args = np.outer(ts, fs)
    M = np.cos(PI2 * args)
    amps = np.linalg.solve(M, ys)
    return amps
```

The first two lines use ts and fs to build the matrix, M. Then np.linalg.solve computes amps.

But there's a hitch. In general we can only solve a system of linear equations if the matrix is square; that is, if the number of equations (rows) is the same as the number of unknowns (columns).

In this example, we have only 4 frequencies, but we evaluated the signal at 11,025 times. So we have many more equations than unknowns.

In general if ys contains more than 4 elements, it is unlikely that we can analyze it using only 4 frequencies.

But in this case, we know that the ys values were actually generated by adding only 4 frequency components, so we can use any 4 values from the wave array to recover amps.

For simplicity, I'll use the first 4 samples from the signal. Using the values of ys, fs, and ts from the previous section, we can run analyze1 like this:

```
n = len(fs)
amps2 = analyze1(ys[:n], fs, ts[:n])
```

And sure enough, amps2 is:

```
[ 0.6   0.25   0.1    0.05 ]
```

This algorithm works, but it is slow. Solving a linear system of equations takes time proportional to n^3, where n is the number of columns in M. We can do better.

Orthogonal Matrices

One way to solve linear systems is by inverting matrices. The inverse of a square matrix M is written M^{-1}, and it has the property that $M^{-1}M = I$. I is the identity matrix, which has the value 1 on all diagonal elements and 0 everywhere else.

So, to solve the equation $y = Ma$, we can multiply both sides by M^{-1}, which yields:

$$M^{-1}y = M^{-1}Ma$$

On the right side, we can replace $M^{-1}M$ with I:

$$M^{-1}y = Ia$$

If we multiply I by any vector a, the result is a, so:

$$M^{-1}y = a$$

This implies that if we can compute M^{-1} efficiently, we can find a with a simple matrix multiplication (using `np.dot`). That takes time proportional to n^2, which is better than n^3.

Inverting a matrix is slow, in general, but some special cases are faster. In particular, if M is **orthogonal**, the inverse of M is just the transpose of M, written M^T. In NumPy transposing an array is a constant-time operation. It doesn't actually move the elements of the array; instead, it creates a "view" that changes the way the elements are accessed.

Again, a matrix is orthogonal if its transpose is also its inverse; that is, $M^T = M^{-1}$. That implies that $M^T M = I$, which means we can check whether a matrix is orthogonal by computing $M^T M$.

So let's see what the matrix looks like in `synthesize2`. In the previous example, M has 11,025 rows, so it might be a good idea to work with a smaller example:

```
def test1():
    amps = np.array([0.6, 0.25, 0.1, 0.05])
    N = 4.0
    time_unit = 0.001
    ts = np.arange(N) / N * time_unit
    max_freq = N / time_unit / 2
    fs = np.arange(N) / N * max_freq
    ys = synthesize2(amps, fs, ts)
```

`amps` is the same vector of amplitudes we saw before. Since we have 4 frequency components, we'll sample the signal at 4 points in time. That way, M is square.

`ts` is a vector of equally spaced sample times in the range from 0 to 1 time unit. I chose the time unit to be 1 millisecond, but it is an arbitrary choice, and we will see in a minute that it drops out of the computation anyway.

Since the frame rate is N samples per time unit, the Nyquist frequency is N / time_unit / 2, which is 2000 Hz in this example. So `fs` is a vector of equally spaced frequencies between 0 and 2000 Hz.

With these values of `ts` and `fs`, the matrix, M, is:

```
[[ 1.    1.     1.     1.   ]
 [ 1.    0.707  0.    -0.707]
 [ 1.    0.    -1.    -0.   ]
 [ 1.   -0.707 -0.     0.707]]
```

You might recognize 0.707 as an approximation of $\sqrt{2}/2$, which is $\cos \pi/4$. You also might notice that this matrix is **symmetric**, which means that the element at (j, k) always equals the element at (k, j). This implies that M is its own transpose; that is, $M^T = M$.

But sadly, M is not orthogonal. If we compute M^TM, we get:

```
[[ 4.   1.  -0.   1.]
 [ 1.   2.   1.  -0.]
 [-0.   1.   2.   1.]
 [ 1.  -0.   1.   2.]]
```

And that's not the identity matrix.

DCT-IV

But if we choose `ts` and `fs` carefully, we can make M orthogonal. There are several ways to do it, which is why there are several versions of the Discrete Cosine Transform (DCT).

One simple option is to shift `ts` and `fs` by a half unit. This version is called DCT-IV, where "IV" is a roman numeral indicating that this is the fourth of eight versions of the DCT.

Here's an updated version of `test1`:

```
def test2():
    amps = np.array([0.6, 0.25, 0.1, 0.05])
    N = 4.0
    ts = (0.5 + np.arange(N)) / N
    fs = (0.5 + np.arange(N)) / 2
    ys = synthesize2(amps, fs, ts)
```

If you compare this to the previous version, you'll notice two changes. First, I added 0.5 to `ts` and `fs`. Second, I canceled out `time_units`, which simplifies the expression for `fs`.

With these values, M is:

```
[[ 0.981  0.831  0.556  0.195]
 [ 0.831 -0.195 -0.981 -0.556]
 [ 0.556 -0.981  0.195  0.831]
 [ 0.195 -0.556  0.831 -0.981]]
```

And M^TM is:

```
[[ 2.   0.   0.   0.]
 [ 0.   2.  -0.   0.]
 [ 0.  -0.   2.  -0.]
 [ 0.   0.  -0.   2.]]
```

Some of the off-diagonal elements are displayed as –0, which means that the floating-point representation is a small negative number. This matrix is very close to $2I$, which means M is almost orthogonal; it's just off by a factor of 2. And for our purposes, that's good enough.

Because M is symmetric and (almost) orthogonal, the inverse of M is just $M/2$. Now we can write a more efficient version of `analyze`:

```
def analyze2(ys, fs, ts):
    args = np.outer(ts, fs)
    M = np.cos(PI2 * args)
    amps = np.dot(M, ys) / 2
    return amps
```

Instead of using `np.linalg.solve`, we just multiply by $M/2$.

Combining `test2` and `analyze2`, we can write an implementation of DCT-IV:

```
def dct_iv(ys):
    N = len(ys)
    ts = (0.5 + np.arange(N)) / N
    fs = (0.5 + np.arange(N)) / 2
    args = np.outer(ts, fs)
    M = np.cos(PI2 * args)
    amps = np.dot(M, ys) / 2
    return amps
```

Again, `ys` is the wave array. We don't have to pass `ts` and `fs` as parameters; `dct_iv` can figure them out based on N, the length of `ys`.

If we've got it right, this function should solve the analysis problem; that is, given `ys` it should be able to recover `amps`. We can test it like this:

```
amps = np.array([0.6, 0.25, 0.1, 0.05])
N = 4.0
ts = (0.5 + np.arange(N)) / N
fs = (0.5 + np.arange(N)) / 2
ys = synthesize2(amps, fs, ts)
amps2 = dct_iv(ys)
max(abs(amps - amps2))
```

Starting with `amps`, we synthesize a wave array, then use `dct_iv` to compute `amps2`. The biggest difference between `amps` and `amps2` is about `1e-16`, which is what we expect due to floating-point errors.

Inverse DCT

Finally, notice that `analyze2` and `synthesize2` are almost identical. The only difference is that `analyze2` divides the result by 2. We can use this insight to compute the inverse DCT:

```
def inverse_dct_iv(amps):
    return dct_iv(amps) * 2
```

`inverse_dct_iv` solves the synthesis problem: it takes the vector of amplitudes and returns the wave array, `ys`. We can test it by starting with `amps`, applying `inverse_dct_iv` and `dct_iv`, and testing that we get back what we started with:

```
amps = [0.6, 0.25, 0.1, 0.05]
ys = inverse_dct_iv(amps)
amps2 = dct_iv(ys)
max(abs(amps - amps2))
```

Again, the biggest difference is about `1e-16`.

The Dct Class

`thinkdsp` provides a `Dct` class that encapsulates the DCT in the same way the `Spectrum` class encapsulates the FFT. To make a `Dct` object, you can invoke `make_dct` on a Wave:

```
signal = thinkdsp.TriangleSignal(freq=400)
wave = signal.make_wave(duration=1.0, framerate=10000)
dct = wave.make_dct()
dct.plot()
```

The result is the DCT of a triangle wave at 400 Hz, shown in Figure 6-2. The values of the DCT can be positive or negative; a negative value in the DCT corresponds to a negated cosine or, equivalently, to a cosine shifted by 180 degrees.

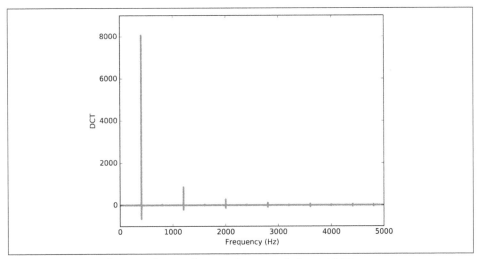

Figure 6-2. DCT of a triangle signal at 400 Hz, sampled at 10 kHz.

`make_dct` uses DCT-II, which is the most common type of DCT, provided by `scipy.fftpack`:

```
import scipy.fftpack

# class Wave:
    def make_dct(self):
        N = len(self.ys)
        hs = scipy.fftpack.dct(self.ys, type=2)
        fs = (0.5 + np.arange(N)) / 2
        return Dct(hs, fs, self.framerate)
```

The results from dct are stored in hs. The corresponding frequencies, computed as in "DCT-IV" on page 71, are stored in fs. And then both are used to initialize the Dct object.

Dct provides make_wave, which performs the inverse DCT. We can test it like this:

```
wave2 = dct.make_wave()
max(abs(wave.ys-wave2.ys))
```

The biggest difference between ys1 and ys2 is about 1e-16, which is what we expect due to floating-point errors.

make_wave uses scipy.fftpack.idct:

```
# class Dct
    def make_wave(self):
        n = len(self.hs)
        ys = scipy.fftpack.idct(self.hs, type=2) / 2 / n
        return Wave(ys, framerate=self.framerate)
```

By default, the inverse DCT doesn't normalize the result, so we have to divide through by *2N*.

Exercises

For the following exercises, I provide some starter code in chap06starter.ipynb. Solutions are in chap06soln.ipynb.

Exercise 6-1.

In this chapter I claim that analyze1 takes time proportional to n^3 and analyze2 takes time proportional to n^2. To see if that's true, run them on a range of input sizes and time them. In Jupyter, you can use the "magic command" %timeit.

If you plot run time versus input size on a log-log scale, you should get a straight line with slope 3 for analyze1 and slope 2 for analyze2.

You also might want to test dct_iv and scipy.fftpack.dct.

Exercise 6-2.

One of the major applications of the DCT is compression for both sound and images. In its simplest form, DCT-based compression works like this:

1. Break a long signal into segments.
2. Compute the DCT of each segment.
3. Identify frequency components with amplitudes so low they are inaudible, and remove them. Store only the frequencies and amplitudes that remain.
4. To play back the signal, load the frequencies and amplitudes for each segment and apply the inverse DCT.

Implement a version of this algorithm and apply it to a recording of music or speech. How many components can you eliminate before the difference is perceptible?

In order to make this method practical, you need some way to store a sparse array; that is, an array where most of the elements are zero. NumPy provides several implementations of sparse arrays, which you can read about at *http://docs.scipy.org/doc/ scipy/reference/sparse.html.*

Exercise 6-3.

In the repository for this book you will find a Jupyter notebook called phase.ipynb that explores the effect of phase on sound perception. Read through this notebook and run the examples. Choose another segment of sound and run the same experiments. Can you find any general relationships between the phase structure of a sound and how we perceive it?

## CHAPTER 7

Discrete Fourier Transform

We've been using the Discrete Fourier Transform (DFT) since Chapter 1, but I haven't explained how it works. Now is the time.

If you understand the Discrete Cosine Transform (DCT), you will understand the DFT. The only difference is that instead of using the cosine function, we'll use the complex exponential function. I'll start by explaining complex exponentials, then we'll follow the same progression as in Chapter 6:

1. We'll start with the synthesis problem: given a set of frequency components and their amplitudes, how can we construct a signal? The synthesis problem is equivalent to the inverse DFT.

2. Then we'll rewrite the synthesis problem in the form of matrix multiplication using NumPy arrays.

3. Next we'll solve the analysis problem, which is equivalent to the DFT: given a signal, how do we find the amplitude and phase offset of its frequency components?

4. Finally, we'll use linear algebra to find a more efficient way to compute the DFT.

The code for this chapter is in `chap07.ipynb`, which is in the repository for this book (see "Using the Code" on page viii). You can also view it at *http://tinyurl.com/thinkdsp07*.

Complex Exponentials

One of the more interesting moves in mathematics is the generalization of an operation from one type to another. For example, a factorial is a function that operates on integers; the natural definition for the factorial of n is the product of all integers from 1 to n.

If you are of a certain inclination, you might wonder how to compute the factorial of a non-integer like 3.5. Since the natural definition doesn't apply, you might look for other ways to compute the factorial function, ways that would work with non-integers.

In 1730, Leonhard Euler found one, a generalization of the factorial function that we know as the gamma function (see *http://en.wikipedia.org/wiki/Gamma_function*).

Euler also found one of the most useful generalizations in applied mathematics, the complex exponential function.

The natural definition of exponentiation is repeated multiplication; for example, $\phi^3 = \phi \cdot \phi \cdot \phi$. But this definition doesn't apply to non-integer exponents.

However, exponentiation can also be expressed as a power series:

$$e^\phi = 1 + \phi + \phi^2/2! + \phi^3/3! + \ldots$$

This definition works with real numbers, with imaginary numbers and, by a simple extension, with complex numbers. Applying this definition to a pure imaginary number, $i\phi$, we get:

$$e^{i\phi} = 1 + i\phi - \phi^2/2! - i\phi^3/3! + \ldots$$

By rearranging terms, we can show that this is equivalent to:

$$e^{i\phi} = \cos\phi + i\sin\phi$$

You can see the derivation at *http://en.wikipedia.org/wiki/Euler's_formula*.

This formula implies that $e^{i\phi}$ is a complex number with magnitude 1; if you think of it as a point in the complex plane, it is always on the unit circle. And if you think of it as a vector, the angle in radians between the vector and the positive x-axis is the argument ϕ.

In the case where the exponent is a complex number, we have:

$$e^{a + i\phi} = e^a e^{i\phi} = A e^{i\phi}$$

where A is a real number that indicates amplitude and $e^{i\phi}$ is a unit complex number that indicates angle.

NumPy provides a version of `exp` that works with complex numbers:

```
>>> phi = 1.5
>>> z = np.exp(1j * phi)
>>> z
(0.0707+0.997j)
```

Python uses `j` to represent the imaginary unit, rather than `i`. A number ending in `j` is considered imaginary, so `1j` is just *i*.

When the argument to `np.exp` is imaginary or complex, the result is a complex number; specifically, an `np.complex128`, which is represented by two 64-bit floating-point numbers. In this example, the result is `0.0707+0.997j`.

Complex numbers have attributes `real` and `imag`:

```
>>> z.real
0.0707
>>> z.imag
0.997
```

To get the magnitude, you can use the built-in function `abs` or `np.absolute`:

```
>>> abs(z)
1.0
>>> np.absolute(z)
1.0
```

To get the angle, you can use `np.angle`:

```
>>> np.angle(z)
1.5
```

This example confirms that $e^{i\phi}$ is a complex number with magnitude 1 and angle ϕ radians.

Complex Signals

If $\phi(t)$ is a function of time, $e^{i\phi(t)}$ is also a function of time. Specifically:

$$e^{i\phi(t)} = \cos \phi(t) + i \sin \phi(t)$$

This function describes a quantity that varies in time, so it is a signal. Specifically, it is a **complex exponential signal**.

In the special case where the frequency of the signal is constant, $\phi(t)$ is $2\pi ft$ and the result is a **complex sinusoid**:

$$e^{i2\pi ft} = \cos 2\pi ft + i \sin 2\pi ft$$

Or more generally, the signal might start at a phase offset ϕ_0, yielding:

$$e^{i\left(2\pi f t + \phi_0\right)}$$

thinkdsp provides an implementation of this signal, ComplexSinusoid:

```
class ComplexSinusoid(Sinusoid):

    def evaluate(self, ts):
        phases = PI2 * self.freq * ts + self.offset
        ys = self.amp * np.exp(1j * phases)
        return ys
```

ComplexSinusoid inherits __init__ from Sinusoid. It provides a version of evalu ate that is almost identical to Sinusoid.evaluate; the only difference is that it uses np.exp instead of np.sin.

The result is a NumPy array of complex numbers:

```
>>> signal = thinkdsp.ComplexSinusoid(freq=1, amp=0.6, offset=1)
>>> wave = signal.make_wave(duration=1, framerate=4)
>>> wave.ys
[ 0.324+0.505j -0.505+0.324j -0.324-0.505j  0.505-0.324j]
```

The frequency of this signal is 1 cycle per second; the amplitude is 0.6 (in unspecified units); and the phase offset is 1 radian.

This example evaluates the signal at four places equally spaced between 0 and 1 second. The resulting samples are complex numbers.

The Synthesis Problem

Just as we did with real sinusoids, we can create compound signals by adding up complex sinusoids with different frequencies. And that brings us to the complex version of the synthesis problem: given the frequency and amplitude of each complex component, how do we evaluate the signal?

The simplest solution is to create ComplexSinusoid objects and add them up:

```
def synthesize1(amps, fs, ts):
    components = [thinkdsp.ComplexSinusoid(freq, amp)
                  for amp, freq in zip(amps, fs)]
    signal = thinkdsp.SumSignal(*components)
    ys = signal.evaluate(ts)
    return ys
```

This function is almost identical to synthesize1 in "Synthesis" on page 66; the only difference is that I replaced CosSignal with ComplexSinusoid.

Here's an example:

```
amps = np.array([0.6, 0.25, 0.1, 0.05])
fs = [100, 200, 300, 400]
framerate = 11025
ts = np.linspace(0, 1, framerate)
ys = synthesize1(amps, fs, ts)
```

The result is:

```
[ 1.000 +0.000e+00j  0.995 +9.093e-02j  0.979 +1.803e-01j ...,
  0.979 -1.803e-01j  0.995 -9.093e-02j  1.000 -5.081e-15j]
```

At the lowest level, a complex signal is a sequence of complex numbers. But how should we interpret it? We have some intuition for real signals: they represent quantities that vary in time; for example, a sound signal represents changes in air pressure. But nothing we measure in the world yields complex numbers.

So what is a complex signal? I don't have a satisfying answer to this question. The best I can offer is two unsatisfying answers:

1. A complex signal is a mathematical abstraction that is useful for computation and analysis, but it does not correspond directly with anything in the real world.

2. If you like, you can think of a complex signal as a sequence of complex numbers that contains two signals as its real and imaginary parts.

Taking the second point of view, we can split the previous signal into its real and imaginary parts:

```
n = 500
thinkplot.plot(ts[:n], ys[:n].real, label='real')
thinkplot.plot(ts[:n], ys[:n].imag, label='imag')
```

Figure 7-1 shows a segment of the result. The real part is a sum of cosines; the imaginary part is a sum of sines. Although the waveforms look different, they contain the same frequency components in the same proportions. To our ears, they sound the same (in general, we don't hear phase offsets).

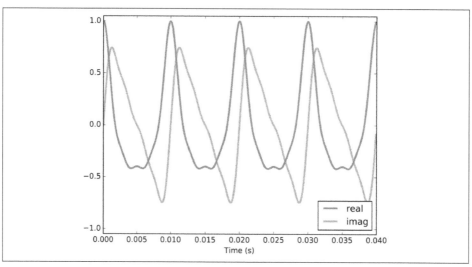

Figure 7-1. Real and imaginary parts of a mixture of complex sinusoids.

Synthesis with Matrices

As we saw in "Synthesis with Arrays" on page 66, we can also express the synthesis problem in terms of matrix multiplication:

```
PI2 = np.pi * 2

def synthesize2(amps, fs, ts):
    args = np.outer(ts, fs)
    M = np.exp(1j * PI2 * args)
    ys = np.dot(M, amps)
    return ys
```

Again, amps is a NumPy array that contains a sequence of amplitudes.

fs is a sequence containing the frequencies of the components. ts contains the times where we will evaluate the signal.

args contains the outer product of ts and fs, with the ts running down the rows and the fs running across the columns (you might want to refer back to Figure 6-1).

Each column of matrix M contains a complex sinusoid with a particular frequency, evaluated at a sequence of ts.

When we multiply M by the amplitudes, the result is a vector whose elements correspond to the ts; each element is the sum of several complex sinusoids, evaluated at a particular time.

Here's the example from the previous section again:

```
>>> ys = synthesize2(amps, fs, ts)
>>> ys
[ 1.000 +0.000e+00j  0.995 +9.093e-02j  0.979 +1.803e-01j ...,
  0.979 -1.803e-01j  0.995 -9.093e-02j  1.000 -5.081e-15j]
```

The result is the same.

In this example the amplitudes are real, but they could also be complex. What effect does a complex amplitude have on the result? Remember that we can think of a complex number in two ways: either the sum of a real and an imaginary part, $x + iy$, or the product of a real amplitude and a complex exponential, $Ae^{i\phi_0}$. Using the second interpretation, we can see what happens when we multiply a complex amplitude by a complex sinusoid. For each frequency, f, we have:

$$Ae^{i\phi_0} \cdot e^{i2\pi ft} = Ae^{i2\pi ft + \phi_0}$$

Multiplying by $Ae^{i\phi_0}$ multiplies the amplitude by A and adds the phase offset ϕ_0.

We can test that claim by running the previous example with complex amplitudes:

```
phi = 1.5
amps2 = amps * np.exp(1j * phi)
ys2 = synthesize2(amps2, fs, ts)

thinkplot.plot(ts[:n], ys.real[:n])
thinkplot.plot(ts[:n], ys2.real[:n])
```

Since amps is an array of reals, multiplying by np.exp(1j * phi) yields an array of complex numbers with phase offset phi radians, and the same magnitudes as amps.

Figure 7-2 shows waveforms with different phase offsets. With $\phi_0 = 1.5$ each frequency component gets shifted by about a quarter of a cycle. But components with different frequencies have different cycles; as a result, each component is shifted by a different amount in time. When we add up the components, the resulting waveforms look different.

Now that we have the more general solution to the synthesis problem—one that handles complex amplitudes—we are ready for the analysis problem.

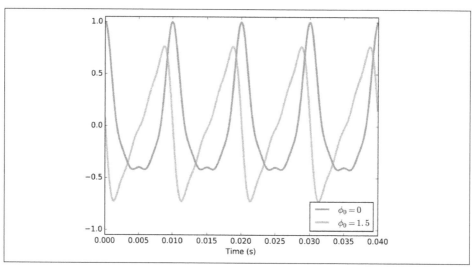

Figure 7-2. Real part of two complex signals that differ by a phase offset.

The Analysis Problem

The analysis problem is the inverse of the synthesis problem: given a sequence of samples, *y*, and knowing the frequencies that make up the signal, can we compute the complex amplitudes of the components, *a*?

As we saw in "Analysis" on page 68, we can solve this problem by forming the synthesis matrix, *M*, and solving the system of linear equations, *Ma = y*, for *a*:

```
def analyze1(ys, fs, ts):
    args = np.outer(ts, fs)
    M = np.exp(1j * PI2 * args)
    amps = np.linalg.solve(M, ys)
    return amps
```

analyze1 takes a (possibly complex) wave array, ys, a sequence of real frequencies, fs, and a sequence of real times, ts. It returns a sequence of complex amplitudes, amps.

Continuing the previous example, we can confirm that analyze1 recovers the amplitudes we started with. For the linear system solver to work, M has to be square, so we need ys, fs and ts, to have the same length. I'll ensure that by slicing ys and ts down to the length of fs:

```
>>> n = len(fs)
>>> amps2 = analyze1(ys[:n], fs, ts[:n])
>>> amps2
[ 0.60+0.j  0.25-0.j  0.10+0.j  0.05-0.j]
```

These are approximately the amplitudes we started with, although each component has a small imaginary part due to floating-point errors.

Efficient Analysis

Unfortunately, solving a linear system of equations is slow. For the DCT, we were able to speed things up by choosing `fs` and `ts` so that M is orthogonal. That way, the inverse of M is the transpose of M, and we can compute both the DCT and inverse DCT by matrix multiplication.

We'll do the same thing for the DFT, with one small change. Since M is complex, we need it to be **unitary**, rather than orthogonal, which means that the inverse of M is the conjugate transpose of M, which we can compute by transposing the matrix and negating the imaginary part of each element. See *http://en.wikipedia.org/wiki/Unitary_matrix*.

The NumPy methods `conj` and `transpose` do what we want. Here's the code that computes M for $N = 4$ components:

```
N = 4
ts = np.arange(N) / N
fs = np.arange(N)
args = np.outer(ts, fs)
M = np.exp(1j * PI2 * args)
```

If M is unitary, $M^*M = I$, where M^* is the conjugate transpose of M, and I is the identity matrix. We can test whether M is unitary like this:

```
MstarM = M.conj().transpose().dot(M)
```

The result, within the tolerance of floating-point error, is $4I$, so M is unitary except for an extra factor of N, similar to the extra factor of 2 we found with the DCT.

We can use this result to write a faster version of `analyze1`:

```
def analyze2(ys, fs, ts):
    args = np.outer(ts, fs)
    M = np.exp(1j * PI2 * args)
    amps = M.conj().transpose().dot(ys) / N
    return amps
```

And test it with appropriate values of `fs` and `ts`:

```
N = 4
amps = np.array([0.6, 0.25, 0.1, 0.05])
fs = np.arange(N)
ts = np.arange(N) / N
ys = synthesize2(amps, fs, ts)
amps3 = analyze2(ys, fs, ts)
```

Again, the result is correct within the tolerance of floating-point arithmetic:

```
[ 0.60+0.j  0.25+0.j  0.10-0.j  0.05-0.j]
```

DFT

As a function, analyze2 would be hard to use because it only works if fs and ts are chosen correctly. Instead, I will rewrite it to take just ys and compute freq and ts itself.

First, I'll make a function to compute the synthesis matrix, M:

```
def synthesis_matrix(N):
    ts = np.arange(N) / N
    fs = np.arange(N)
    args = np.outer(ts, fs)
    M = np.exp(1j * PI2 * args)
    return M
```

Then I'll write the function that takes ys and returns amps:

```
def analyze3(ys):
    N = len(ys)
    M = synthesis_matrix(N)
    amps = M.conj().transpose().dot(ys) / N
    return amps
```

We are almost done; analyze3 computes something very close to the DFT, with one difference. The conventional definition of the DFT does not divide by N:

```
def dft(ys):
    N = len(ys)
    M = synthesis_matrix(N)
    amps = M.conj().transpose().dot(ys)
    return amps
```

Now we can confirm that my version yields the same result as np.fft.fft:

```
>>> dft(ys)
[ 2.4+0.j  1.0+0.j  0.4-0.j  0.2-0.j]
```

The result is close to amps * N. And here's the version in np.fft:

```
>>> np.fft.fft(ys)
[ 2.4+0.j  1.0+0.j  0.4-0.j  0.2-0.j]
```

They are the same, within floating-point error.

The inverse DFT is almost the same, except we don't have to transpose and conjugate M, and *now* we have to divide through by N:

```
def idft(ys):
    N = len(ys)
    M = synthesis_matrix(N)
    amps = M.dot(ys) / N
    return amps
```

Finally, we can confirm that dft(idft(amps)) yields amps:

```
>>> ys = idft(amps)
>>> dft(ys)
[ 0.60+0.j  0.25+0.j  0.10-0.j  0.05-0.j]
```

If I could go back in time, I might change the definition of the DFT so it divides by N and the inverse DFT doesn't. That would be more consistent with my presentation of the synthesis and analysis problems.

Or I might change the definition so that both operations divide through by $\sqrt{N}$. Then the DFT and inverse DFT would be more symmetric.

But I can't go back in time (yet!), so we're stuck with a slightly weird convention. For practical purposes it doesn't really matter.

The DFT Is Periodic

In this chapter I presented the DFT in the form of matrix multiplication. We compute the synthesis matrix, M, and the analysis matrix, M^*. When we multiply M^* by the wave array, y, each element of the result is the product of a row from M^* and y, which we can write in the form of a summation:

$$\mathrm{DFT}(y)[k] = \sum_n y[n] \exp\left(-2\pi ink/N\right)$$

where k is an index of frequency from 0 to $N - 1$ and n is an index of time from 0 to $N - 1$. So $\mathrm{DFT}(y)[k]$ is the kth element of the DFT of y.

Normally we evaluate this summation for N values of k, running from 0 to $N - 1$. We *could* evaluate it for other values of k, but there is no point, because they start to repeat. That is, the value at k is the same as the value at $k + N$ or $k + 2N$ or $k - N$, etc.

We can see that mathematically by plugging $k + N$ into the summation:

$$\mathrm{DFT}(y)[k + N] = \sum_n y[n] \exp\left(-2\pi in(k + N)/N\right)$$

Since there is a sum in the exponent, we can break it into two parts:

$$\mathrm{DFT}(y)[k + N] = \sum_n y[n] \exp\left(-2\pi ink/N\right) \exp\left(-2\pi inN/N\right)$$

In the second term, the exponent is always an integer multiple of 2π, so the result is always 1, and we can drop it:

$$\mathrm{DFT}(y)[k + N] = \sum_n y[n] \exp\left(-2\pi ink/N\right)$$

And we can see that this summation is equivalent to DFT$(y)[k]$. So the DFT is periodic, with period N. You will need this result for one of the exercises at the end of this chapter, which asks you to implement the Fast Fourier Transform (FFT).

As an aside, writing the DFT in the form of a summation provides an insight into how it works. If you review the diagram in "Synthesis with Arrays" on page 66, you'll see that each column of the synthesis matrix is a signal evaluated at a sequence of times. The analysis matrix is the (conjugate) transpose of the synthesis matrix, so each *row* is a signal evaluated at a sequence of times.

Therefore, each summation is the correlation of y with one of the signals in the array (see "Correlation as Dot Product" on page 61). That is, each element of the DFT is a correlation that quantifies the similarity of the wave array, y, and a complex exponential at a particular frequency.

DFT of Real Signals

The Spectrum class in thinkdsp is based on np.ftt.rfft, which computes the "real DFT"; that is, it works with real signals. But the DFT as presented in this chapter is more general than that; it works with complex signals.

So what happens when we apply the "full DFT" to a real signal? Let's look at an example:

```
signal = thinkdsp.SawtoothSignal(freq=500)
wave = signal.make_wave(duration=0.1, framerate=10000)
hs = dft(wave.ys)
amps = np.absolute(hs)
```

This code makes a sawtooth wave with frequency 500 Hz, sampled at frame rate 10 kHz. hs contains the complex DFT of the wave; amps contains the amplitude at each frequency. But what frequency do these amplitudes correspond to? If we look at the body of dft, we see:

```
fs = np.arange(N)
```

It's tempting to think that these values are the right frequencies. The problem is that dft doesn't know the sampling rate. The DFT assumes that the duration of the wave is 1 time unit, so it thinks the sampling rate is N per time unit. In order to interpret the frequencies, we have to convert from these arbitrary time units back to seconds, like this:

```
fs = np.arange(N) * framerate / N
```

With this change, the range of frequencies is from 0 to the actual frame rate, 10 kHz. Now we can plot the spectrum:

```
thinkplot.plot(fs, amps)
thinkplot.config(xlabel='frequency (Hz)',
                 ylabel='amplitude')
```

Figure 7-3 shows the amplitude of the signal for each frequency component from 0 to 10 kHz. The left half of the figure is what we should expect: the dominant frequency is at 500 Hz, with harmonics dropping off like $1/f$.

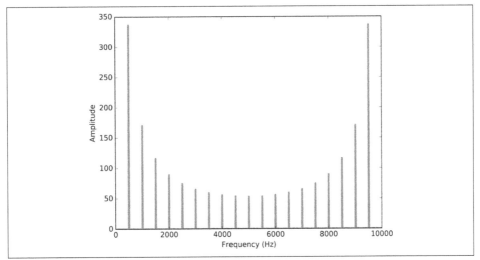

Figure 7-3. DFT of a 500 Hz sawtooth signal sampled at 10 kHz.

But the right half of the figure is a surprise. Past 5000 Hz, the amplitude of the harmonics starts growing again, peaking at 9500 Hz. What's going on?

The answer: aliasing. Remember that with frame rate 10,000 Hz, the folding frequency is 5000 Hz. As we saw in "Aliasing" on page 17, a component at 5500 Hz is indistinguishable from a component at 4500 Hz. When we evaluate the DFT at 5500 Hz, we get the same value as at 4500 Hz. Similarly, the value at 6000 Hz is the same as the one at 4000 Hz, and so on.

The DFT of a real signal is symmetric around the folding frequency. Since there is no additional information past this point, we can save time by evaluating only the first half of the DFT, and that's exactly what `np.fft.rfft` does.

Exercises

Solutions to these exercises are in `chap07soln.ipynb`.

Exercise 7-1.

The notebook for this chapter, `chap07.ipynb`, contains additional examples and explanations. Read through it and run the code.

Exercise 7-2.

In this chapter, I showed how we can express the DFT and inverse DFT as matrix multiplications. These operations take time proportional to N^2, where N is the length of the wave array. That is fast enough for many applications, but there is a faster algorithm, the Fast Fourier Transform (FFT), which takes time proportional to $N \log N$.

The key to the FFT is the Danielson–Lanczos lemma:

$$\mathrm{DFT}(y)[n] = \mathrm{DFT}(e)[n] + \exp(-2\pi i n / N)\mathrm{DFT}(o)[n]$$

where $\mathrm{DFT}(y)[n]$ is the *n*th element of the DFT of y, e is a wave array containing the even elements of y, and o contains the odd elements of y.

This lemma suggests a recursive algorithm for the DFT:

1. Given a wave array, y, split it into its even elements, e, and its odd elements, o.
2. Compute the DFT of e and o by making recursive calls.
3. Compute DFT(y) for each value of n using the Danielson–Lanczos lemma.

For the base case of this recursion, you could wait until the length of y is 1. In that case, DFT(y) = y. Or if the length of y is sufficiently small, you could compute its DFT by matrix multiplication, possibly using a precomputed matrix.

Hint: I suggest you implement this algorithm incrementally by starting with a version that is not truly recursive. In Step 2, instead of making a recursive call, use `dft`, as defined in "DFT" on page 86, or `np.fft.fft`. Get Step 3 working, and confirm that the results are consistent with the other implementations. Then add a base case and confirm that it works. Finally, replace Step 2 with recursive calls.

One more hint: remember that the DFT is periodic; you might find `np.tile` useful.

You can read more about the FFT at *https://en.wikipedia.org/wiki/Fast_Fourier_trans form*.

## CHAPTER 8

Filtering and Convolution

In this chapter I present one of the most important and useful ideas related to signal processing: the Convolution Theorem. But before we can understand the Convolution Theorem, we have to understand convolution. I'll start with a simple example, smoothing, and we'll go from there.

The code for this chapter is in `chap08.ipynb`, which is in the repository for this book (see "Using the Code" on page viii). You can also view it at *http://tinyurl.com/thinkdsp08*.

Smoothing

Smoothing is an operation that tries to remove short-term variations from a signal in order to reveal long-term trends. For example, if you were to plot daily changes in the price of a stock, it would look noisy; a smoothing operator might make it easier to see whether the price was generally going up or down over time.

A common smoothing algorithm is a moving average, which computes the mean of the previous n values, for some value of n.

For example, Figure 8-1 shows the daily closing price of Facebook stock from May 17, 2012 to December 8, 2015. The gray line is the raw data, and the darker line shows the 30-day moving average. Smoothing removes the most extreme changes and makes it easier to see long-term trends.

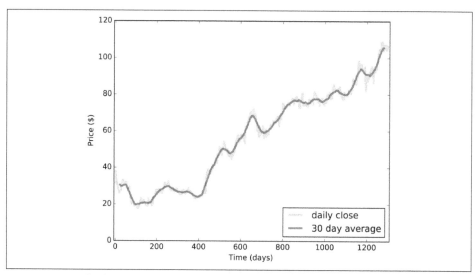

Figure 8-1. Daily closing price of Facebook stock and a 30-day moving average.

Smoothing operations also apply to sound signals. As an example, I'll start with a square wave at 440 Hz. As we saw in "Square Waves" on page 16, the harmonics of a square wave drop off slowly, so it contains many high-frequency components.

First I'll construct the signal and two waves:

```
signal = thinkdsp.SquareSignal(freq=440)
wave = signal.make_wave(duration=1, framerate=44100)
segment = wave.segment(duration=0.01)
```

wave is a one-second slice of the signal; segment is a shorter slice I'll use for plotting.

To compute the moving average of this signal, I'll use a window similar to the ones in "Windowing" on page 33. Previously we used a Hamming window to avoid spectral leakage caused by discontinuity at the beginning and end of a signal. More generally, we can use windows to compute the weighted sum of samples in a wave.

For example, to compute a moving average, I'll create a window with 11 elements and normalize it so the elements add up to 1:

```
window = np.ones(11)
window /= sum(window)
```

Now I can compute the average of the first 11 elements by multiplying the window by the wave array:

```
ys = segment.ys
N = len(ys)
padded = thinkdsp.zero_pad(window, N)
prod = padded * ys
sum(prod)
```

padded is a version of the window with zeros added to the end so it is the same length as segment.ys. Adding zeros like this is called **padding**.

prod is the product of the window and the wave array. The sum of the elementwise products is the average of the first 11 elements of the array. Since these elements are all –1, their average is –1.

To compute the next element of the moving average, we **roll** the window, which shifts the ones to the right and wraps one of the zeros from the end around to the beginning.

When we multiply the rolled window and the wave array, we get the average of the next 11 elements of the wave array, starting with the second:

```
rolled = np.roll(rolled, 1)
prod = rolled * ys
sum(prod)
```

The result is –1 again.

We can compute the rest of the elements the same way. The following function wraps the code we have seen so far in a loop and stores the results in an array:

```
def smooth(ys, window):
    N = len(ys)
    smoothed = np.zeros(N)
    padded = thinkdsp.zero_pad(window, N)
    rolled = padded

    for i in range(N):
        smoothed[i] = sum(rolled * ys)
        rolled = np.roll(rolled, 1)
    return smoothed
```

smoothed is the array that will contain the results; padded is an array that contains the window and enough zeros to have length N; and rolled is a copy of padded that gets shifted to the right by one element each time through the loop.

Inside the loop, we multiply ys by rolled to select 11 elements and add them up.

Figure 8-2 shows the result for a square wave. The gray line is the original signal; the darker line is the smoothed signal. The smoothed signal starts to ramp up when the leading edge of the window reaches the first transition, and levels off when the window crosses the transition. As a result, the transitions are less abrupt, and the corners less sharp. If you listen to the smoothed signal, it sounds less buzzy and slightly muffled.

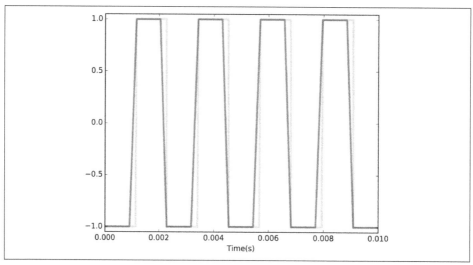

Figure 8-2. A square signal at 400 Hz (gray) and an 11-element moving average.

Convolution

The operation we just performed—applying a window function to each overlapping segment of a wave—is called **convolution**.

Convolution is such a common operation that NumPy provides an implementation that is simpler and faster than my version:

```
convolved = np.convolve(ys, window, mode='valid')
smooth2 = thinkdsp.Wave(convolved, framerate=wave.framerate)
```

`np.convolve` computes the convolution of the wave array and the window. The mode flag `valid` indicates that it should only compute values when the window and the wave array overlap completely, so it stops when the right edge of the window reaches the end of the wave array. Other than that, the result is the same as in Figure 8-2.

Actually, there is one other difference. The loop in the previous section actually computes **cross-correlation**:

$$(f \star g)[n] = \sum_{m=0}^{N-1} f[m]g[n + m]$$

where f is a wave array with length N, g is the window, and $\star$ is the symbol for cross-correlation. To compute the nth element of the result, we shift g to the right, which is why the index is $n + m$.

The definition of convolution is slightly different:

$$(f * g)[n] = \sum_{m=0}^{N-1} f[m]g[n-m]$$

The symbol * represents convolution. The difference is in the index of g: m has been negated, so the summation iterates the elements of g backward (assuming that negative indices wrap around to the end of the array).

Because the window we used in this example is symmetric, cross-correlation and convolution yield the same result. When we use other windows, we will have to be more careful.

You might wonder why convolution is defined like this, with the window applied in a way that seems backward. There are two reasons:

- This definition comes up naturally for several applications, especially analysis of signal-processing systems, which is the topic of Chapter 10.
- Also, this definition is the basis of the Convolution Theorem, coming up very soon.

Finally, a note for people who know too much: in the presentation so far I have not distinguished between convolution and circular convolution. We'll get to it.

The Frequency Domain

Smoothing makes the transitions in a square signal less abrupt, and makes the sound slightly muffled. Let's see what effect this operation has on the spectrum. First I'll plot the spectrum of the original wave:

```
spectrum = wave.make_spectrum()
spectrum.plot(color=GRAY)
```

Then the smoothed wave:

```
convolved = np.convolve(wave.ys, window, mode='same')
smooth = thinkdsp.Wave(convolved, framerate=wave.framerate)
spectrum2 = smooth.make_spectrum()
spectrum2.plot()
```

The mode flag same indicates that the result should have the same length as the input. In this example, it will include a few values that "wrap around", but that's OK for now.

Figure 8-3 shows the result. The fundamental frequency is almost unchanged; the first few harmonics are attenuated, and the higher harmonics are almost eliminated. So smoothing has the effect of a low-pass filter, which we saw in "Spectrums" on page 7 and "Pink Noise" on page 46.

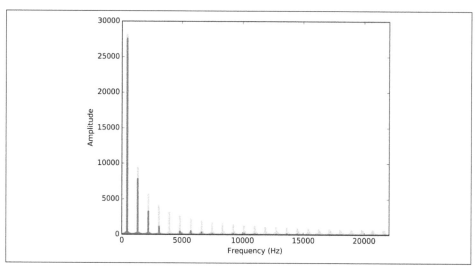

Figure 8-3. Spectrum of the square wave before and after smoothing.

To see how much each component has been attenuated, we can compute the ratio of the two spectrums:

```
amps = spectrum.amps
amps2 = spectrum2.amps
ratio = amps2 / amps
ratio[amps<560] = 0
thinkplot.plot(ratio)
```

`ratio` is the ratio of the amplitude before and after smoothing. When `amps` is small, this ratio can be big and noisy, so for simplicity I set the ratio to 0 except where the harmonics are.

Figure 8-4 shows the result. As expected, the ratio is high for low frequencies and drops off at a cutoff frequency near 4000 Hz. But there is another feature we did not expect: above the cutoff, the ratio bounces around between 0 and 0.2. What's up with that?

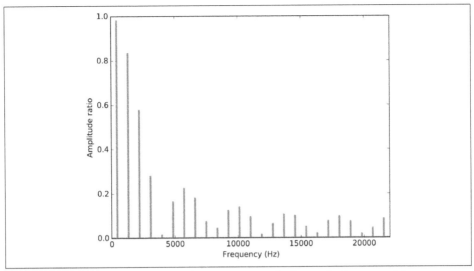

Figure 8-4. Ratio of spectrums for the square wave, before and after smoothing.

The Convolution Theorem

The answer is the Convolution Theorem. Stated mathematically:

$$\mathrm{DFT}(f * g) = \mathrm{DFT}(f) \cdot \mathrm{DFT}(g)$$

where f is a wave array and g is a window. In words, the Convolution Theorem says that if we convolve f and g, and then compute the DFT, we get the same answer as when computing the DFT of f and g, and then multiplying the results elementwise.

When we apply an operation like convolution to a wave, we say we are working in the **time domain**, because the wave is a function of time. When we apply an operation like multiplication to the DFT, we are working in the **frequency domain**, because the DFT is a function of frequency.

Using these terms, we can state the Convolution Theorem more concisely:

> Convolution in the time domain corresponds to multiplication in the frequency domain.

And that explains Figure 8-4, because when we convolve a wave and a window, we multiply the spectrum of the wave with the spectrum of the window. To see how that works, we can compute the DFT of the window:

```
padded = zero_pad(window, N)
dft_window = np.fft.rfft(padded)
thinkplot.plot(abs(dft_window))
```

padded contains the smoothing window, padded with zeros to be the same length as wave; dft_window contains the DFT of padded.

Figure 8-5 shows the result, along with the ratios we computed in the previous section. The ratios are exactly the amplitudes in dft_window. Mathematically:

$$\text{abs}(\text{DFT}(f * g))/\text{abs}(\text{DFT}(f)) = \text{abs}(\text{DFT}(g))$$

In this context, the DFT of a window is called a **filter**. For any convolution window in the time domain, there is a corresponding filter in the frequency domain. And for any filter that can be expressed by elementwise multiplication in the frequency domain, there is a corresponding window.

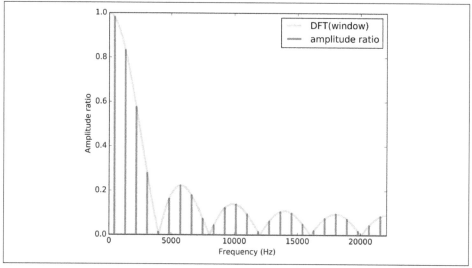

Figure 8-5. Ratio of spectrums for the square wave, before and after smoothing, along with the DFT of the smoothing window.

Gaussian Filter

The moving average window we used in the previous section is a low-pass filter, but it is not a very good one. The DFT drops off steeply at first, but then it bounces around. Those bounces are called **sidelobes**, and they are there because the moving average window is like a square wave, so its spectrum contains high-frequency harmonics that drop off proportionally to $1/f$, which is relatively slow.

We can do better with a Gaussian window. SciPy provides functions that compute many common convolution windows, including `gaussian`:

```
gaussian = scipy.signal.gaussian(M=11, std=2)
gaussian /= sum(gaussian)
```

`M` is the number of elements in the window; `std` is the standard deviation of the Gaussian distribution used to compute it. Figure 8-6 shows the shape of the window. It is a discrete approximation of the Gaussian "bell curve". The figure also shows the moving average window from the previous example, which is sometimes called a **boxcar window** because it looks like a rectangular railway car.

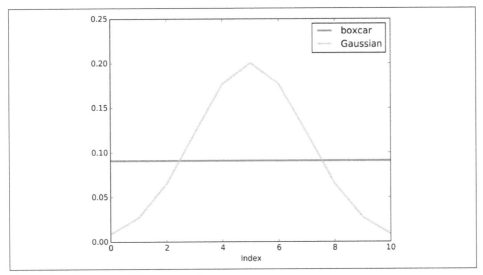

Figure 8-6. Boxcar and Gaussian windows.

I ran the computations from the previous sections again with this window and generated Figure 8-7, which shows the ratio of the spectrums before and after smoothing, along with the DFT of the Gaussian window.

As a low-pass filter, Gaussian smoothing is better than a simple moving average. After the ratio drops off, it stays low, with almost none of the sidelobes we saw with the boxcar window. So it does a better job of cutting off the higher frequencies.

The reason it does so well is that the DFT of a Gaussian curve is also a Gaussian curve. So the ratio drops off in proportion to $\exp\left(-f^2\right)$, which is much faster than $1/f$.

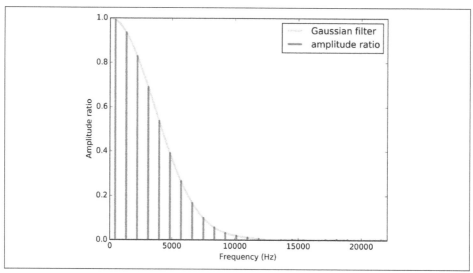

Figure 8-7. Ratio of spectrums before and after Gaussian smoothing, and the DFT of the window.

Efficient Convolution

One of the reasons the FFT is such an important algorithm is that, combined with the Convolution Theorem, it provides an efficient way to compute convolution, cross-correlation, and autocorrelation.

Again, the Convolution Theorem states:

$$\text{DFT}(f * g) = \text{DFT}(f) \cdot \text{DFT}(g)$$

So one way to compute a convolution is:

$$f * g = \text{IDFT}(\text{DFT}(f) \cdot \text{DFT}(g))$$

where IDFT is the inverse DFT. A simple implementation of convolution takes time proportional to N^2; this algorithm, using the FFT, takes time proportional to $N \log N$.

We can confirm that it works by computing the same convolution both ways. As an example, I'll apply it to the Facebook stock data shown in Figure 8-1:

```
import pandas as pd

names = ['date', 'open', 'high', 'low', 'close', 'volume']
df = pd.read_csv('fb.csv', header=0, names=names)
ys = df.close.values[::-1]
```

This example uses Pandas to read the data from the CSV file (included in the repository for this book). If you are not familiar with Pandas, don't worry: I'm not going to do much with it in this book. But if you're interested, you can learn more about it in *Think Stats* at *http://thinkstats2.com*.

The result, df, is a DataFrame, one of the data structures provided by Pandas. close is a NumPy array that contains daily closing prices.

Next I'll create a Gaussian window and convolve it with close:

```
window = scipy.signal.gaussian(M=30, std=6)
window /= window.sum()
smoothed = np.convolve(ys, window, mode='valid')
```

fft_convolve computes the same thing using the FFT:

```
from np.fft import fft, ifft

def fft_convolve(signal, window):
    fft_signal = fft(signal)
    fft_window = fft(window)
    return ifft(fft_signal * fft_window)
```

We can test it by padding the window to the same length as ys and then computing the convolution:

```
padded = zero_pad(window, N)
smoothed2 = fft_convolve(ys, padded)
```

The result has $M - 1$ bogus values at the beginning, where M is the length of the window. We can slice off the bogus values like this:

```
M = len(window)
smoothed2 = smoothed2[M-1:]
```

The result agrees with fft_convolve with about 12 digits of precision.

Efficient Autocorrelation

In "Convolution" on page 94 I presented definitions of cross-correlation and convolution, and we saw that they are almost the same, except that in convolution the window is reversed.

Now that we have an efficient algorithm for convolution, we can also use it to compute cross-correlations and autocorrelations. Using the data from the previous section, we can compute the autocorrelation of Facebook stock prices:

```
corrs = np.correlate(close, close, mode='same')
```

With mode='same', the result has the same length as close, corresponding to lags from $-N/2$ to $N/2 - 1$. The gray line in Figure 8-8 shows the result. Except at lag=0, there are no peaks, so there is no apparent periodic behavior in this signal. However, the autocorrelation function drops off slowly, suggesting that this signal resembles pink noise, as we saw in "Autocorrelation" on page 57.

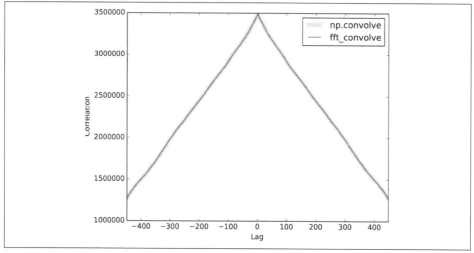

Figure 8-8. Autocorrelation functions computed by NumPy and fft_autocorr.

To compute autocorrelation using convolution, we have to zero-pad the signal to double the length. This trick is necessary because the FFT is based on the assumption that the signal is periodic; that is, that it wraps around from the end to the beginning. With time-series data like this, that assumption is invalid. Adding zeros, and then trimming the results, removes the bogus values.

Also, remember that convolution reverses the direction of the window. In order to cancel that effect, we reverse the direction of the window before calling fft_con volve, using np.flipud, which flips a NumPy array. The result is a view of the array, not a copy, so this operation is fast:

```
def fft_autocorr(signal):
    N = len(signal)
    signal = thinkdsp.zero_pad(signal, 2*N)
    window = np.flipud(signal)

    corrs = fft_convolve(signal, window)
    corrs = np.roll(corrs, N//2+1)[:N]
    return corrs
```

The result from `fft_convolve` has length $2N$. Of those, the first and last $N/2$ are valid; the rest are the result of zero-padding. To select the valid element, we roll the results and select the first N, corresponding to lags from $-N/2$ to $N/2 - 1$.

As shown in Figure 8-8, the results from `fft_autocorr` and `np.correlate` are identical (with about 9 digits of precision).

Notice that the correlations in Figure 8-8 are large numbers; we could normalize them (between –1 and 1) as shown in "Using NumPy" on page 62.

The strategy we used here for autocorrelation also works for cross-correlation. Again, you have to prepare the signals by flipping one and padding both, and then you have to trim the invalid parts of the result. This padding and trimming is a nuisance, but that's why libraries like NumPy provide functions to do it for you.

Exercises

Solutions to these exercises are in `chap08soln.ipynb`.

Exercise 8-1.

The notebook for this chapter is `chap08.ipynb`. Read through it and run the code.

It contains an interactive widget that lets you experiment with the parameters of the Gaussian window to see what effect they have on the cutoff frequency.

What goes wrong when you increase the width of the Gaussian window, `std`, without increasing the number of elements in the window, `M`?

Exercise 8-2.

In this chapter I claimed that the Fourier Transform of a Gaussian curve is also a Gaussian curve. For Discrete Fourier Transforms, this relationship is approximately true.

Try it out for a few examples. What happens to the Fourier Transform as you vary `std`?

Exercise 8-3.

If you did the exercises in Chapter 3, you saw the effect of the Hamming window, and some of the other windows provided by NumPy, on spectral leakage. We can get some insight into the effects of these windows by looking at their DFTs.

In addition to the Gaussian window we used in this chapter, create a Hamming window with the same size. Zero-pad the windows and plot their DFTs. Which window acts as a better low-pass filter? You might find it useful to plot the DFTs on a log-y scale.

Experiment with a few different windows and a few different sizes.

## CHAPTER 9

Differentiation and Integration

This chapter picks up where the previous chapter left off, looking at the relationship between windows in the time domain and filters in the frequency domain.

In particular, we'll look at the effects of a finite difference window, which approximates differentiation, and the cumulative sum operation, which approximates integration.

The code for this chapter is in chap09.ipynb, which is in the repository for this book (see "Using the Code" on page viii). You can also view it at *http://tinyurl.com/thinkdsp09*.

Finite Differences

In "Smoothing" on page 91, we applied a smoothing window to the daily stock price of Facebook and found that a smoothing window in the time domain corresponds to a low-pass filter in the frequency domain.

In this section, we'll look at daily price changes and see that computing the difference between successive elements, in the time domain, corresponds to a high-pass filter.

Here's the code to read the data, store it as a wave, and compute its spectrum:

```
import pandas as pd

names = ['date', 'open', 'high', 'low', 'close', 'volume']
df = pd.read_csv('fb.csv', header=0, names=names)
ys = df.close.values[::-1]
close = thinkdsp.Wave(ys, framerate=1)
spectrum = wave.make_spectrum()
```

This example uses Pandas to read the CSV file; the result is a `DataFrame`, `df`, with columns for the opening price, closing price, and high and low prices. I select the closing prices and save them in a `Wave` object. The frame rate is 1 sample per day.

Figure 9-1 shows this time series and its spectrum. Visually, the time series resembles Brownian noise (see "Brownian Noise" on page 43). And the spectrum looks like a straight line, albeit a noisy one. The estimated slope is −1.9, which is consistent with Brownian noise.

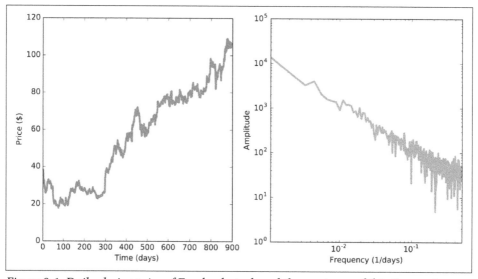

Figure 9-1. Daily closing price of Facebook stock and the spectrum of this time series.

Now let's compute the daily price change using `np.diff`:

```
diff = np.diff(ys)
change = thinkdsp.Wave(diff, framerate=1)
change_spectrum = change.make_spectrum()
```

Figure 9-2 shows the resulting wave and its spectrum. The daily changes resemble white noise, and the estimated slope of the spectrum, −0.06, is near zero, which is what we expect for white noise.

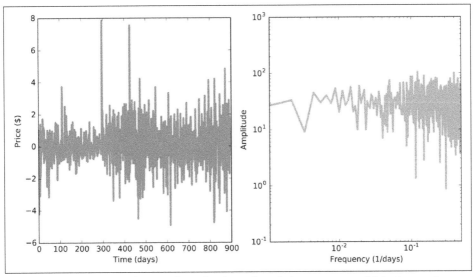

Figure 9-2. Daily price change of Facebook stock and the spectrum of this time series.

The Frequency Domain

Computing the difference between successive elements is the same as convolution with the window [1, -1]. If the order of those elements seems backward, remember that convolution reverses the window before applying it to the signal.

We can see the effect of this operation in the frequency domain by computing the DFT of the window:

```
diff_window = np.array([1.0, -1.0])
padded = thinkdsp.zero_pad(diff_window, len(close))
diff_wave = thinkdsp.Wave(padded, framerate=close.framerate)
diff_filter = diff_wave.make_spectrum()
```

Figure 9-3 shows the result. The finite difference window corresponds to a high-pass filter: its amplitude increases with frequency, linearly for low frequencies and then sublinearly after that. In the next section, we'll see why.

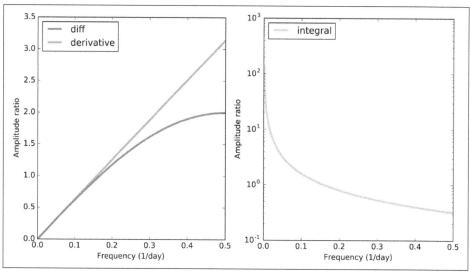

Figure 9-3. Filters corresponding to the diff and differentiate operators (left) and integration operator (right, log-y scale).

Differentiation

The window we used in the previous section is a numerical approximation of the first derivative, so the filter approximates the effect of differentiation.

Differentiation in the time domain corresponds to a simple filter in the frequency domain; we can figure out what it is with a little math.

Suppose we have a complex sinusoid with frequency f:

$$E_f(t) = e^{2\pi i f t}$$

The first derivative of E_f is:

$$\frac{d}{dt}E_f(t) = 2\pi i f e^{2\pi i f t}$$

which we can rewrite as:

$$\frac{d}{dt}E_f(t) = 2\pi i f E_f(t)$$

In other words, taking the derivative of E_f is the same as multiplying by $2\pi i f$, which is a complex number with magnitude $2\pi f$ and angle $\pi/2$.

We can compute the filter that corresponds to differentiation like this:

```
deriv_filter = close.make_spectrum()
deriv_filter.hs = PI2 * 1j * deriv_filter.fs
```

I started with the spectrum of `close`, which has the right size and frame rate, then replaced the hs with $2\pi i f$. Figure 9-3 (left) shows this filter; it is a straight line.

As we saw in "Synthesis with Matrices" on page 82, multiplying a complex sinusoid by a complex number has two effects: it multiplies the amplitude, in this case by $2\pi f$, and shifts the phase offset, in this case by $\pi/2$.

If you are familiar with the language of operators and eigenfunctions, each E_f is an eigenfunction of the differentiation operator, with the corresponding eigenvalue $2\pi i f$. See *http://en.wikipedia.org/wiki/Eigenfunction*.

If you are not familiar with that language, here's what it means:

- An operator is a function that takes a function and returns another function. For example, differentiation is an operator.

- A function, g, is an eigenfunction of an operator, $\mathscr{A}$, if applying $\mathscr{A}$ to g has the effect of multiplying the function by a scalar. That is, $\mathscr{A}g = \lambda g$.

- In that case, the scalar λ is the eigenvalue that corresponds to the eigenfunction g.

- A given operator might have many eigenfunctions, each with a corresponding eigenvalue.

Because complex sinusoids are eigenfunctions of the differentiation operator, they are easy to differentiate. All we have to do is multiply by a complex scalar.

For signals with more than one component, the process is only slightly harder:

1. Express the signal as the sum of complex sinusoids.

2. Compute the derivative of each component by multiplication.

3. Add up the differentiated components.

If that process sounds familiar, that's because it is identical to the algorithm for convolution in "Efficient Convolution" on page 100: compute the DFT, multiply by a filter, and compute the inverse DFT.

Spectrum provides a method that applies the differentiation filter:

```
# class Spectrum:

    def differentiate(self):
        self.hs *= PI2 * 1j * self.fs
```

We can use it to compute the derivative of the Facebook time series:

```
deriv_spectrum = close.make_spectrum()
deriv_spectrum.differentiate()
deriv = deriv_spectrum.make_wave()
```

Figure 9-4 compares the daily price changes computed by `np.diff` with the derivative we just computed. I selected the first 50 values in the time series so we can see the differences more clearly.

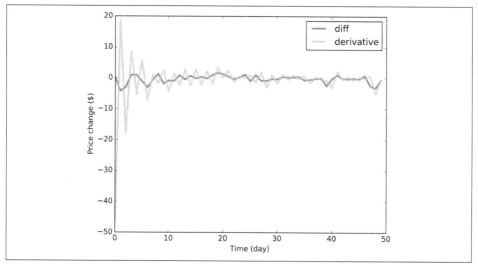

Figure 9-4. Comparison of daily price changes computed by np.diff and by applying the differentiation filter.

The derivative is noisier, because it amplifies the high-frequency components more, as shown in Figure 9-3 (left). Also, the first few elements of the derivative are very noisy. The problem there is that the DFT-based derivative is based on the assumption that the signal is periodic. In effect, it connects the last element in the time series back to the first element, which creates artifacts at the boundaries.

To summarize, we have shown:

- Computing the difference between successive values in a signal can be expressed as convolution with a simple window. The result is an approximation of the first derivative.

- Differentiation in the time domain corresponds to a simple filter in the frequency domain. For periodic signals, the result is the first derivative, exactly. For some non-periodic signals, it can approximate the derivative.

Using the DFT to compute derivatives is the basis of **spectral methods** for solving differential equations (see *http://en.wikipedia.org/wiki/Spectral_method*).

In particular, it is useful for the analysis of linear, time-invariant systems, which is coming up in Chapter 10.

Integration

In the previous section, we showed that differentiation in the time domain corresponds to a simple filter in the frequency domain: it multiplies each component by $2\pi i f$. Since integration is the inverse of differentiation, it also corresponds to a simple filter: it divides each component by $2\pi i f$.

We can compute this filter like this:

```
integ_filter = close.make_spectrum()
integ_filter.hs = 1 / (PI2 * 1j * integ_filter.fs)
```

Figure 9-3 (right) shows this filter on a log-y scale, which makes it easier to see.

Spectrum provides a method that applies the integration filter:

```
# class Spectrum:

    def integrate(self):
        self.hs /= PI2 * 1j * self.fs
```

We can confirm that the integration filter is correct by applying it to the spectrum of the derivative we just computed:

```
integ_spectrum = deriv_spectrum.copy()
integ_spectrum.integrate()
```

But notice that at $f = 0$, we are dividing by 0. The result in NumPy is NaN, which is a special floating-point value that represents "not a number". We can partially deal with this problem by setting this value to 0 before converting the spectrum back to a wave:

```
integ_spectrum.hs[0] = 0
integ_wave = integ_spectrum.make_wave()
```

Figure 9-5 shows this integrated derivative along with the original time series. They are almost identical, but the integrated derivative has been shifted down. The problem is that when we clobbered the $f = 0$ component, we set the bias of the signal to 0. But that should not be surprising; in general, differentiation loses information about the bias, and integration can't recover it. In some sense, the NaN at $f = 0$ is telling us that this element is unknown.

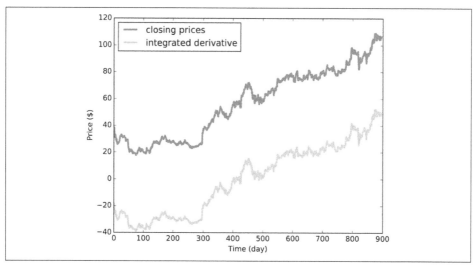

Figure 9-5. Comparison of the original time series and the integrated derivative.

If we provide this "constant of integration", the results are identical, which confirms that this integration filter is the correct inverse of the differentiation filter.

Cumulative Sum

In the same way that the `diff` operator approximates differentiation, the cumulative sum approximates integration. I'll demonstrate with a sawtooth signal:

```
signal = thinkdsp.SawtoothSignal(freq=50)
in_wave = signal.make_wave(duration=0.1, framerate=44100)
```

Figure 9-6 shows this wave and its spectrum.

`Wave` provides a method that computes the cumulative sum of a wave array and returns a new `Wave` object:

```
# class Wave:

    def cumsum(self):
        ys = np.cumsum(self.ys)
        ts = self.ts.copy()
        return Wave(ys, ts, self.framerate)
```

We can use it to compute the cumulative sum of `in_wave`:

```
out_wave = in_wave.cumsum()
out_wave.unbias()
```

Figure 9-7 shows the resulting wave and its spectrum. If you did the exercises in Chapter 2, this waveform should look familiar: it's a parabolic signal.

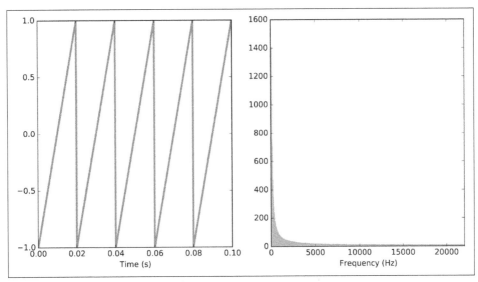

Figure 9-6. A sawtooth wave and its spectrum.

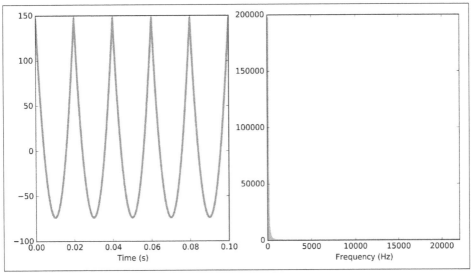

Figure 9-7. A parabolic wave and its spectrum.

Comparing the two, we see that the amplitudes of the components drop off more quickly in the spectrum of the parabolic signal than in the spectrum of the sawtooth signal. In Chapter 2, we saw that the components of the sawtooth drop off in proportion to $1/f$. Since the cumulative sum approximates integration, and integration filters components in proportion to $1/f$, the components of the parabolic wave drop off in proportion to $1/f^2$.

We can see that graphically by computing the filter that corresponds to the cumulative sum:

```
cumsum_filter = diff_filter.copy()
cumsum_filter.hs = 1 / cumsum_filter.hs
```

Because cumsum is the inverse operation of diff, we start with a copy of diff_filter, which is the filter that corresponds to the diff operation, and then invert the hs.

Figure 9-8 shows the filters corresponding to cumulative sum and integration. The cumulative sum is a good approximation of integration except at the highest frequencies, where it drops off a little faster.

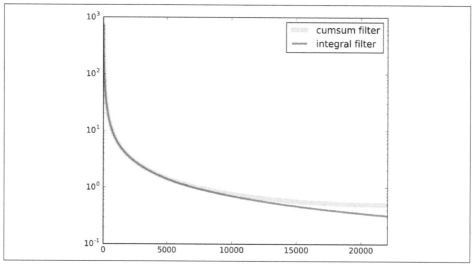

Figure 9-8. Filters corresponding to cumulative sum and integration.

To confirm that this is the correct filter for the cumulative sum, we can compare it to the ratio of the spectrum out_wave to the spectrum of in_wave:

```
in_spectrum = in_wave.make_spectrum()
out_spectrum = out_wave.make_spectrum()
ratio_spectrum = out_spectrum.ratio(in_spectrum, thresh=1)
```

And here's the method that computes the ratios:

```
def ratio(self, denom, thresh=1):
    ratio_spectrum = self.copy()
    ratio_spectrum.hs /= denom.hs
    ratio_spectrum.hs[denom.amps < thresh] = np.nan
    return ratio_spectrum
```

When denom.amps is small, the resulting ratio is noisy, so I set those values to NaN.

Figure 9-9 shows the ratios and the filter corresponding to the cumulative sum. They agree, which confirms that inverting the filter for `diff` yields the filter for `cumsum`.

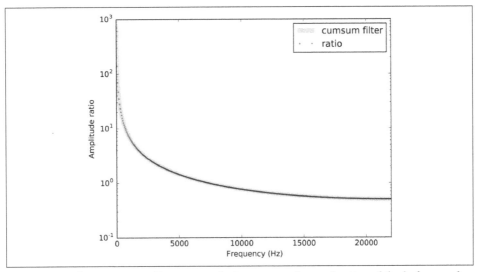

Figure 9-9. Filter corresponding to cumulative sum and actual ratios of the before-and-after spectrums.

Finally, we can confirm that the Convolution Theorem applies by applying the `cum sum` filter in the frequency domain:

```
out_wave2 = (in_spectrum * cumsum_filter).make_wave()
```

Within the limits of floating-point error, `out_wave2` is identical to `out_wave`, which we computed using `cumsum`, so the Convolution Theorem works! But note that this demonstration only works with periodic signals.

Integrating Noise

In "Brownian Noise" on page 43, we generated Brownian noise by computing the cumulative sum of white noise. Now that we understand the effect of `cumsum` in the frequency domain, we have some insight into the spectrum of Brownian noise.

White noise has equal power at all frequencies, on average. When we compute the cumulative sum, the amplitude of each component is divided by f. Since power is the square of magnitude, the power of each component is divided by f^2. So on average, the power at frequency f is proportional to $1/f^2$:

$$P_f = K/f^2$$

where K is a constant that's not important. Taking the log of both sides yields:

$$\log P_f = \log K - 2 \log f$$

And that's why, when we plot the spectrum of Brownian noise on a log-log scale, we expect to see a straight line with slope −2, at least approximately.

In "Finite Differences" on page 105 we plotted the spectrum of closing prices for Facebook stock, and estimated that the slope is −1.9, which is consistent with Brownian noise. Many stock prices have similar spectrums.

When we used the diff operator to compute daily changes, we multiplied the *amplitude* of each component by a filter proportional to f, which means we multiplied the *power* of each component by f^2. On a log-log scale, this operation adds 2 to the slope of the power spectrum, which is why the estimated slope of the result is near 0.1 (but a little lower, because diff only approximates differentiation).

Exercises

Solutions to these exercises are in chap09soln.ipynb.

Exercise 9-1.

The notebook for this chapter is chap09.ipynb. Read through it and run the code.

In "Cumulative Sum" on page 112, I mentioned that some of the examples don't work with non-periodic signals. Try replacing the sawtooth wave, which is periodic, with the Facebook data, which is not, and see what goes wrong.

Exercise 9-2.

The goal of this exercise is to explore the effects of diff and differentiate on a signal. Create a triangle wave and plot it. Apply diff and plot the result. Compute the spectrum of the triangle wave, apply differentiate, and plot the result. Convert the spectrum back to a wave and plot it. Are there differences between the effects of diff and differentiate for this wave?

Exercise 9-3.

The goal of this exercise is to explore the effects of cumsum and integrate on a signal. Create a square wave and plot it. Apply cumsum and plot the result. Compute the spectrum of the square wave, apply integrate, and plot the result. Convert the spectrum back to a wave and plot it. Are there differences between the effects of cumsum and integrate for this wave?

Exercise 9-4.

The goal of this exercise is to explore the effect of integrating twice. Create a sawtooth wave, compute its spectrum, then apply integrate twice. Plot the resulting wave and its spectrum. What is the mathematical form of the wave? Why does it resemble a sinusoid?

Exercise 9-5.

The goal of this exercise is to explore the effects of the second difference and second derivative. Create a CubicSignal, which is defined in thinkdsp. Compute the second difference by applying diff twice. What does the result look like? Compute the second derivative by applying differentiate to the spectrum twice. Does the result look the same?

Plot the filters that correspond to the second difference and the second derivative and compare them. Hint: in order to get the filters on the same scale, use a wave with frame rate 1.

## CHAPTER 10

LTI Systems

This chapter presents the theory of signals and systems, using musical acoustics as an example. It explains an important application of the Convolution Theorem, characterization of linear, time-invariant systems (which I'll define soon).

The code for this chapter is in `chap10.ipynb`, which is in the repository for this book (see "Using the Code" on page viii). You can also view it at *http://tinyurl.com/thinkdsp10*.

Signals and Systems

In the context of signal processing, a **system** is an abstract representation of anything that takes a signal as input and produces a signal as output.

For example, an electronic amplifier is a circuit that takes an electrical signal as input and produces a (louder) signal as output.

As another example, when you listen to a musical performance, you can think of the room as a system that takes the sound of the performance at the location where it is generated and produces a somewhat different sound at the location where you hear it.

A **linear, time-invariant system**[1] is a system with these two properties:

Linearity
> If you put two inputs into the system at the same time, the result is the sum of their outputs. Mathematically, if an input x_1 produces output y_1 and another input x_2 produces y_2, then $ax_1 + bx_2$ produces $ay_1 + by_2$, where a and b are scalars.

---

[1] My presentation here follows *http://en.wikipedia.org/wiki/LTI_system_theory*.

Time invariance

The effect of the system doesn't vary over time, or depend on the state of the system. So if inputs x_1 and x_2 differ only by a shift in time, their outputs, y_1 and y_2, differ by the same shift but are otherwise identical.

Many physical systems have these properties, at least approximately:

- Circuits that contain only resistors, capacitors, and inductors are LTI, to the degree that the components behave like their idealized models.
- Mechanical systems that contain springs, masses, and dashpots are also LTI, assuming linear springs (force proportional to displacement) and dashpots (force proportional to velocity).
- Also, and most relevant to applications in this book, the media that transmit sound (including air, water, and solids) are well modeled by LTI systems.

LTI systems are described by linear differential equations, and the solutions of those equations are complex sinusoids (see *http://en.wikipedia.org/wiki/Linear_differen tial_equation*).

This result provides an algorithm for computing the effect of an LTI system on an input signal:

1. Express the signal as the sum of complex sinusoid components.
2. For each input component, compute the corresponding output component.
3. Add up the output components.

At this point, I hope this algorithm sounds familiar. It's the same algorithm we used for convolution in "Efficient Convolution" on page 100, and for differentiation in "Differentiation" on page 108. This process is called **spectral decomposition** because we "decompose" the input signal into its spectral components.

In order to apply this process to an LTI system, we have to **characterize** the system by finding its effect on each component of the input signal. For mechanical systems, it turns out that there is a simple and efficient way to do that: you kick it and record the output.

Technically, the "kick" is called an **impulse** and the output is called the **impulse response**. You might wonder how a single impulse can completely characterize a system. You can see the answer by computing the DFT of an impulse. Here's a wave array with an impulse at $t = 0$:

```
impulse = np.zeros(8)
impulse[0] = 1
impulse_spectrum = np.fft.fft(impulse)
```

Here's the wave array:

```
[ 1.  0.  0.  0.  0.  0.  0.  0.]
```

And here's its spectrum:

```
[ 1.+0.j  1.+0.j  1.+0.j  1.+0.j  1.+0.j  1.+0.j  1.+0.j  1.+0.j]
```

The spectrum is all ones; that is, an impulse is the sum of components with equal magnitudes at all frequencies. This spectrum should not be confused with white noise, which has the same *average* power at all frequencies, but varies around that average.

When you test a system by inputting an impulse, you are testing the response of the system at all frequencies. And you can test them all at the same time because the system is linear, so simultaneous tests don't interfere with each other.

Windows and Filters

To show why this kind of system characterization works, I will start with a simple example: a 2-element moving average. We can think of this operation as a system that takes a signal as an input and produces a slightly smoother signal as an output.

In this example we know what the window is, so we can compute the corresponding filter. But that's not usually the case; in the next section we'll look at an example where we don't know the window or the filter ahead of time.

Here's a window that computes a 2-element moving average (see "Smoothing" on page 91):

```
window_array = np.array([0.5, 0.5, 0, 0, 0, 0, 0, 0,])
window = thinkdsp.Wave(window_array, framerate=8)
```

We can find the corresponding filter by computing the DFT of the window:

```
filtr = window.make_spectrum(full=True)
```

Figure 10-1 shows the result. The filter that corresponds to a moving average window is a low-pass filter with the approximate shape of a Gaussian curve.

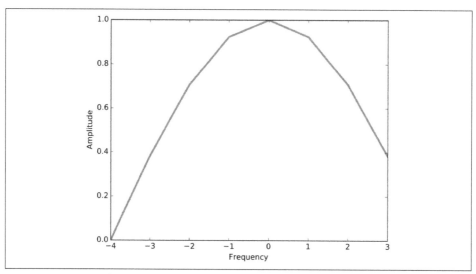

Figure 10-1. DFT of a 2-element moving average window.

Now imagine that we did not know the window or the corresponding filter, and we wanted to characterize this system. We would do that by inputting an impulse and measuring the impulse response.

In this example, we can compute the impulse response by multiplying the spectrum of the impulse and the filter, and then converting the result from a spectrum to a wave:

```
product = impulse_spectrum * filtr
filtered = product.make_wave()
```

Since `impulse_spectrum` is all ones, the product is identical to the filter, and the filtered wave is identical to the window.

This example demonstrates two things:

- Because the spectrum of an impulse is all ones, the DFT of the impulse response is identical to the filter that characterizes the system.

- Therefore, the impulse response is identical to the convolution window that characterizes the system.

Acoustic Response

To characterize the acoustic response of a room or open space, a simple way to generate an impulse is to pop a balloon or fire a gun. The result is an input signal that approximates an impulse, so the sound you hear approximates the impulse response.

As an example, I'll use a recording of a gunshot to characterize the room where the gun was fired, then use the impulse response to simulate the effect of that room on a violin recording.

This example is in chap10.ipynb, which is in the repository for this book; you can also view it, and listen to the examples, at *http://tinyurl.com/thinkdsp10*.

Here's the gunshot:

```
response = thinkdsp.read_wave('180961__kleeb__gunshots.wav')
response = response.segment(start=0.26, duration=5.0)
response.normalize()
response.plot()
```

I select a segment starting at 0.26 seconds to remove the silence before the gunshot. Figure 10-2 (left) shows the waveform of the gunshot. Next I compute the DFT of response:

```
transfer = response.make_spectrum()
transfer.plot()
```

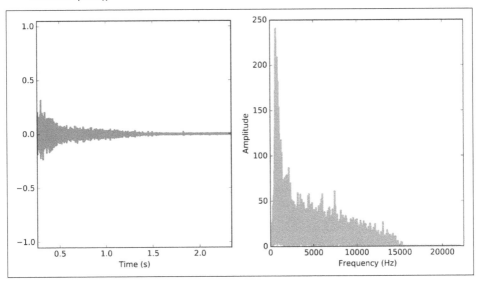

Figure 10-2. Waveform of a gunshot.

Figure 10-2 (right) shows the result. This spectrum encodes the response of the room; for each frequency, the spectrum contains a complex number that represents an amplitude multiplier and a phase shift. This spectrum is called a **transfer function** because it contains information about how the system transfers the input to the output.

Now we can simulate the effect this room would have on the sound of a violin. Here is the violin recording we used in "Periodic Signals" on page 1:

```
violin = thinkdsp.read_wave('92002__jcveliz__violin-origional.wav')
violin.truncate(len(response))
violin.normalize()
```

The violin and gunshot waves were sampled at the same frame rate, 44,100 Hz. And coincidentally, the duration of both is about the same. I trimmed the violin wave to the same length as the gunshot.

Next I compute the DFT of the violin wave:

```
spectrum = violin.make_spectrum()
```

Now I know the magnitude and phase of each frequency component in the input, and I know the transfer function of the system. Their product is the DFT of the output, which we can use to compute the output wave:

```
output = (spectrum * transfer).make_wave()
output.normalize()
output.plot()
```

Figure 10-3 shows the input (top) and output (bottom) of the system. They are substantially different, and the differences are clearly audible. Load chap10.ipynb and listen to them. One thing I find striking about this example is that you can get a sense of what the room is like; to me, it sounds like a long, narrow room with hard floors and ceilings. That is, like a firing range.

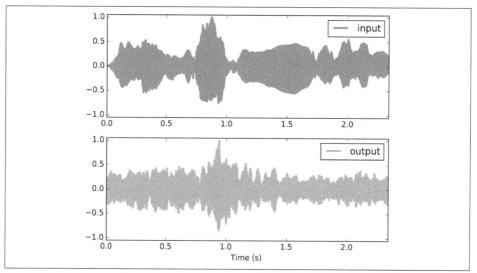

Figure 10-3. The waveform of the violin recording before and after convolution.

There's one thing I glossed over in this example that I'll mention in case it bothers anyone. The violin recording I started with has already been transformed by one system: the room where it was recorded. So what I really computed in my example is the sound of the violin after two transformations. To properly simulate the sound of a violin in a different room, I should have characterized the room where the violin was recorded and applied the inverse of that transfer function first.

Systems and Convolution

If you think the previous example is black magic, you are not alone. I've been thinking about it for a while and it still makes my head hurt.

In the previous section, I suggested one way to think about it:

- An impulse is made up of components with amplitude 1 at all frequencies.
- The impulse response contains the sum of the responses of the system to all of these components.
- The transfer function, which is the DFT of the impulse response, encodes the effect of the system on each frequency component in the form of an amplitude multiplier and a phase shift.
- For any input, we can compute the response of the system by breaking the input into components, computing the response to each component, and adding them up.

But if you don't like that, there's another way to think about it altogether: convolution! By the Convolution Theorem, multiplication in the frequency domain corresponds to convolution in the time domain. In this example, the output of the system is the convolution of the input and the system response.

Here are the keys to understanding why that works:

- You can think of the samples in the input wave as a sequence of impulses with varying amplitude.
- Each impulse in the input yields a copy of the impulse response, shifted in time (because the system is time-invariant) and scaled by the amplitude of the input.
- The output is the sum of the shifted, scaled copies of the impulse response. The copies add up because the system is linear.

Let's work our way up gradually. Suppose that instead of firing one gun, we fire two: a big one with amplitude 1 at $t = 0$ and a smaller one with amplitude 0.5 at $t = 1$.

We can compute the response of the system by adding up the original impulse response and a scaled, shifted copy of itself. Here's a function that makes a shifted, scaled copy of a wave:

```
def shifted_scaled(wave, shift, factor):
    res = wave.copy()
    res.shift(shift)
    res.scale(factor)
    return res
```

The parameter `shift` is a time shift in seconds; `factor` is a multiplicative factor.

Here's how we use it to compute the response to a two-gun salute:

```
shift = 1
factor = 0.5
gun2 = response + shifted_scaled(response, shift, factor)
```

Figure 10-4 shows the result. You can hear what it sounds like in `chap10.ipynb`. Not surprisingly, it sounds like two gunshots, the first one louder than the second.

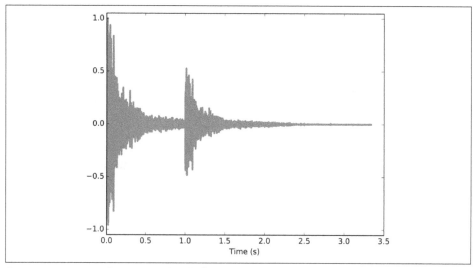

Figure 10-4. Sum of a wave and a shifted, scaled copy.

Now suppose instead of 2 guns, you were to add up 100 guns fired at a rate of 441 shots per second. This loop computes the result:

```
dt = 1 / 441
total = 0
for k in range(100):
    total += shifted_scaled(response, k*dt, 1.0)
```

With 441 shots per second, you don't hear the individual shots. Instead, it sounds like a periodic signal at 441 Hz. If you play this example, it sounds like a car horn in a garage.

And that brings us to a key insight: you can think of any wave as a series of samples, where each sample is an impulse with a different amplitude.

As an example, I'll generate a sawtooth signal at 441 Hz:

```
signal = thinkdsp.SawtoothSignal(freq=441)
wave = signal.make_wave(duration=0.1,
                        framerate=response.framerate)
```

Now I'll loop through the series of impulses that make up the sawtooth, and add up the impulse responses:

```
total = 0
for t, y in zip(wave.ts, wave.ys):
    total += shifted_scaled(response, t, y)
```

The result is what it would sound like to play a sawtooth wave in a firing range. Again, you can listen to it in chap10.ipynb.

Figure 10-5 shows a diagram of this computation, where f is the sawtooth, g is the impulse response, and h is the sum of the shifted, scaled copies of g.

$$
\begin{array}{llllllll}
f[0] & [& g[0] & g[1] & g[2] & \cdots & &] \\
f[1] & [& & g[0] & g[1] & g[2] & \cdots &] \\
f[2] & [& & & g[0] & g[1] & g[2] & \cdots &] \\
\\
& [& & & h[2] & & &]
\end{array}
$$

Figure 10-5. Diagram of the sum of scaled and shifted copies of g.

For the example shown:

$$h[2] = f[0]g[2] + f[1]g[1] + f[2]g[0]$$

And more generally:

$$h[n] = \sum_{m=0}^{N-1} f[m]g[n-m]$$

You might recognize this equation from "Convolution" on page 94. It is the convolution of f and g. This shows that if the input is f and the impulse response of the system is g, the output is the convolution of f and g.

In summary, there are two ways to think about the effect of a system on a signal:

1. The input is a sequence of impulses, so the output is the sum of scaled, shifted copies of the impulse response; that sum is the convolution of the input and the impulse response.

2. The DFT of the impulse response is a transfer function that encodes the effect of the system on each frequency component as a magnitude and phase offset. The DFT of the input encodes the magnitude and phase offset of the frequency components it contains. Multiplying the DFT of the input by the transfer function yields the DFT of the output.

The equivalence of these descriptions should not be a surprise. It is basically a statement of the Convolution Theorem: convolution of f and g in the time domain corresponds to multiplication in the frequency domain.

And if you wondered why convolution is defined as it is, which seemed backward when we talked about smoothing and difference windows, now you know the reason: the definition of convolution appears naturally in the response of an LTI system to a signal.

Proof of the Convolution Theorem

Well, I've put it off long enough. It's time to prove the Convolution Theorem (CT), which states:

$$\mathrm{DFT}(f * g) = \mathrm{DFT}(f)\mathrm{DFT}(g)$$

where f and g are vectors with the same length, N.

I'll proceed in two steps:

1. I'll show that in the special case where f is a complex exponential, convolution with g has the effect of multiplying f by a scalar.
2. In the more general case where f is not a complex exponential, we can use the DFT to express it as a sum of exponential components, compute the convolution of each component (by multiplication), and then add up the results.

Together these steps prove the Convolution Theorem. But first, let's assemble the pieces we'll need. The DFT of g, which I'll call G, is:

$$\mathrm{DFT}(g)[k] = G[k] = \sum_n g[n] \exp\left(-2\pi i n k / N\right)$$

where k is an index of frequency from 0 to $N - 1$ and n is an index of time from 0 to $N - 1$. The result is a vector of N complex numbers.

The inverse DFT of F, which I'll call f, is:

$$\mathrm{IDFT}(F)[n] = f[n] = \sum_k F[k] \exp\left(2\pi i n k / N\right)$$

Here's the definition of convolution:

$$(f * g)[n] = \sum_m f[m]g[n - m]$$

where m is another index of time from 0 to $N - 1$. Convolution is commutative, so I could equivalently write:

$$(f * g)[n] = \sum_m f[n - m]g[m]$$

Now let's consider the special case where f is a complex exponential with frequency k, which I'll call e_k:

$$f[n] = e_k[n] = \exp(2\pi i n k/N)$$

where k is an index of frequency and n is an index of time.

Plugging e_k into the second definition of convolution yields:

$$\left(e_k * g\right)[n] = \sum_m \exp(2\pi i(n - m)k/N)g[m]$$

We can split the first term into a product:

$$\left(e_k * g\right)[n] = \sum_m \exp(2\pi i n k/N) \exp(- 2\pi i m k/N)g[m]$$

The first half does not depend on m, so we can pull it out of the summation:

$$\left(e_k * g\right)[n] = \exp(2\pi i n k/N)\sum_m \exp(- 2\pi i m k/N)g[m]$$

Now we recognize that the first term is e_k, and the summation is $G[k]$ (using m as the index of time). So we can write:

$$\left(e_k * g\right)[n] = e_k[n]G[k]$$

which shows that for each complex exponential, e_k, convolution with g has the effect of multiplying e_k by $G[k]$. In mathematical terms, each e_k is an eigenvector of this operation, and $G[k]$ is the corresponding eigenvalue (see "Differentiation" on page 108).

Now for the second part of the proof. If the input signal, f, doesn't happen to be a complex exponential, we can express it as a sum of complex exponentials by computing its DFT, F. For each value of k from 0 to $N-1$, $F[k]$ is the complex magnitude of the component with frequency k.

Each input component is a complex exponential with magnitude $F[k]$, so each output component is a complex exponential with magnitude $F[k]G[k]$, based on the first part of the proof.

Because the system is linear, the output is just the sum of the output components:

$$(f * g)[n] = \sum_k F[k]G[k]e_k[n]$$

Plugging in the definition of e_k yields:

$$(f * g)[n] = \sum_k F[k]G[k] \exp(2\pi i n k / N)$$

The righthand side is the inverse DFT of the product FG. Thus:

$$(f * g) = \mathrm{IDFT}(FG)$$

Substituting $F = \mathrm{DFT}(f)$ and $G = \mathrm{DFT}(g)$:

$$(f * g) = \mathrm{IDFT}(\mathrm{DFT}(f)\mathrm{DFT}(g))$$

Finally, taking the DFT of both sides yields the Convolution Theorem:

$$\mathrm{DFT}(f * g) = \mathrm{DFT}(f)\mathrm{DFT}(g)$$

QED.

Exercises

Solutions to these exercises are in `chap10soln.ipynb`.

Exercise 10-1.

In "Systems and Convolution" on page 125 I describe convolution as the sum of shifted, scaled copies of a signal.

But in "Acoustic Response" on page 122, when we multiply the DFT of the signal by the transfer function, that operation corresponds to **circular convolution**, which assumes that the signal is periodic. As a result, you might notice that the output contains an extra note at the beginning, which wraps around from the end.

Fortunately, there is a standard solution to this problem. If you add enough zeros to the end of the signal before computing the DFT, you can avoid the wraparound effect.

Modify the example in chap10.ipynb and confirm that zero-padding eliminates the extra note at the beginning of the output.

Exercise 10-2.

The Open AIR library provides a "centralized... on-line resource for anyone interested in auralization and acoustical impulse response data" (*http://www.openairlib.net*). Browse its collection of impulse response data and download one that sounds interesting. Find a short recording that has the same sample rate as the impulse response you downloaded.

Simulate the sound of your recording in the space where the impulse response was measured, computed two ways: by convolving the recording with the impulse response and by computing the filter that corresponds to the impulse response and multiplying by the DFT of the recording.

## CHAPTER 11

Modulation and Sampling

In "Aliasing" on page 17 we saw that when a signal is sampled at 10,000 Hz, a component at 5500 Hz is indistinguishable from a component at 4500 Hz. In this example, the folding frequency, 5000 Hz, is half of the sampling rate. But I didn't explain why.

This chapter explores the effect of sampling and presents the Sampling Theorem, which explains aliasing and the folding frequency.

I'll start by exploring the effect of convolution with impulses; I'll use that effect to explain amplitude modulation (AM), which turns out to be useful for understanding the Sampling Theorem.

The code for this chapter is in `chap11.ipynb`, which is in the repository for this book (see "Using the Code" on page viii). You can also view it at *http://tinyurl.com/ thinkdsp-ch11*.

Convolution with Impulses

As we saw in "Systems and Convolution" on page 125, convolution of a signal with a series of impulses has the effect of adding up shifted, scaled copies of the signal.

As an example, I'll read a signal that sounds like a beep:

```
filename = '253887__themusicalnomad__positive-beeps.wav'
wave = thinkdsp.read_wave(filename)
wave.normalize()
```

And I'll construct a wave with four impulses:

```
imp_sig = thinkdsp.Impulses([0.005, 0.3, 0.6,  0.9],
                      amps=[1,     0.5, 0.25, 0.1])
impulses = imp_sig.make_wave(start=0, duration=1.0,
                          framerate=wave.framerate)
```

and then convolve them:

```
convolved = wave.convolve(impulses)
```

Figure 11-1 shows the results, with the signal in the top left, the impulses in the lower left, and the result on the right.

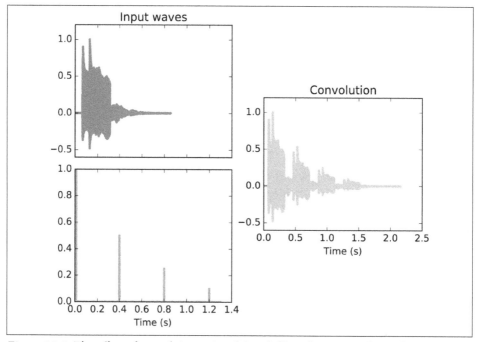

Figure 11-1. The effect of convolving a signal (top left) with a series of impulses (bottom left). The result (right) is the sum of shifted, scaled copies of the signal.

You can hear the result in `chap11.ipynb`; it sounds like a series of four beeps with decreasing loudness.

The point of this example is just to demonstrate that convolution with impulses makes shifted, scaled copies. This result will be useful in the next section.

Amplitude Modulation

Amplitude modulation (AM) is used to broadcast AM radio, among other applications. At the transmitter, the signal (which might contain speech, music, etc.) is "modulated" by multiplying it with a cosine signal that acts as a "carrier wave". The result is a high-frequency wave that is suitable for broadcast by radio. Typical frequencies for AM radio in the United States are 500–1600 kHz (see *https://en.wikipedia.org/wiki/AM_broadcasting*).

At the receiving end, the broadcast signal is "demodulated" to recover the original signal. Surprisingly, demodulation works by multiplying the broadcast signal, again, by the same carrier wave.

To see how that works, I'll modulate a signal with a carrier wave at 10 kHz. Here's the signal:

```
filename = '105977__wcfl10__favorite-station.wav'
wave = thinkdsp.read_wave(filename)
wave.unbias()
wave.normalize()
```

And here's the carrier:

```
carrier_sig = thinkdsp.CosSignal(freq=10000)
carrier_wave = carrier_sig.make_wave(duration=wave.duration,
                            framerate=wave.framerate)
```

We can multiply them using the * operator, which multiplies the wave arrays elementwise:

```
modulated = wave * carrier_wave
```

The result sounds pretty bad. You can hear it in chap11.ipynb.

Figure 11-2 shows what's happening in the frequency domain. The top row is the spectrum of the original signal. The next row is the spectrum of the modulated signal, after multiplying by the carrier. It contains two copies of the original spectrum, shifted by plus and minus 10 kHz.

To understand why, recall that convolution in the time domain corresponds to multiplication in the frequency domain. Conversely, multiplication in the time domain corresponds to convolution in the frequency domain. When we multiply the signal by the carrier, we are convolving its spectrum with the DFT of the carrier.

Since the carrier is a simple cosine wave, its DFT is two impulses, at plus and minus 10 kHz. Convolution with these impulses makes shifted, scaled copies of the spectrum. Notice that the amplitude of the spectrum is smaller after modulation. The energy from the original signal is split between the copies.

We demodulate the signal by multiplying by the carrier wave again:

```
demodulated = modulated * carrier_wave
```

The third row of Figure 11-2 shows the result. Again, multiplication in the time domain corresponds to convolution in the frequency domain, which makes shifted, scaled copies of the spectrum.

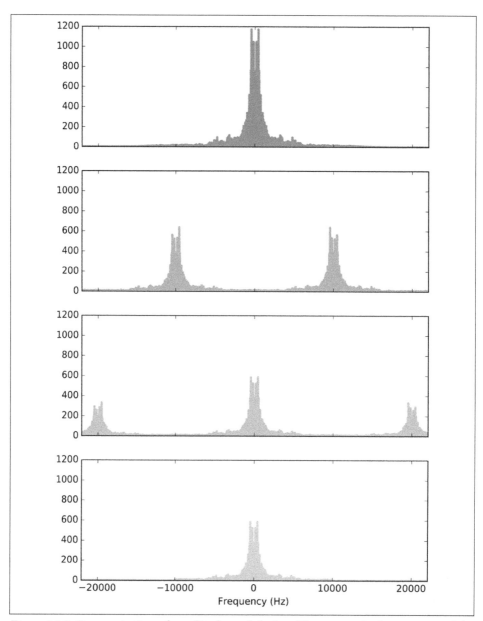

Figure 11-2. Demonstration of amplitude modulation. The top row is the spectrum of the signal; the next row is the spectrum after modulation; the next row is the spectrum after demodulation; the last row is the demodulated signal after low-pass filtering.

Since the modulated spectrum contains two peaks, each peak gets split in half and shifted by plus and minus 20 kHz. Two of the copies meet at 0 kHz and get added together; the other two copies end up centered at plus and minus 20 kHz.

If you listen to the demodulated signal, it sounds pretty good. The extra copies of the spectrum add high-frequency components that were not in the original signal. These are so high that most speakers can't play them and most people can't hear them, but if you have good speakers and good ears, you might.

In that case, you can get rid of the extra components by applying a low-pass filter:

```
demodulated_spectrum = demodulated.make_spectrum(full=True)
demodulated_spectrum.low_pass(10000)
filtered = demodulated_spectrum.make_wave()
```

The result is quite close to the original wave, although about half of the power is lost after demodulating and filtering. That's not a problem in practice, because much more of the power is lost in transmitting and receiving the broadcast signal. Since we have to amplify the result anyway, another factor of 2 is not an issue.

Sampling

I explained amplitude modulation in part because it is interesting, but mostly because it will help us understand sampling. **Sampling** is the process of measuring an analog signal at a series of points in time, usually with equal spacing.

For example, the WAV files we have used as examples were recorded by sampling the output of a microphone using an analog-to-digital converter (ADC). The sampling rate for most of them is 44.1 kHz, which is the standard rate for "CD-quality" sound, or 48 kHz, which is the standard for DVD sound.

At 48 kHz, the folding frequency is 24 kHz, which is higher than most people can hear (see *https://en.wikipedia.org/wiki/Hearing_range*).

In most of these waves, each sample has 16 bits, so there are 2^{16} distinct levels. This "bit depth" turns out to be enough that adding more bits does not improve the sound quality noticeably (see *https://en.wikipedia.org/wiki/Digital_audio*).

Of course, applications other than audio signals might require higher sampling rates in order to capture higher frequencies, or higher bit depth in order to reproduce waveforms with more fidelity.

To demonstrate the effect of the sampling process, I am going to start with a wave sampled at 44.1 kHz and select samples from it at about 11 kHz. This is not exactly the same as sampling from an analog signal, but the effect is the same.

First I'll load a recording of a drum solo:

```
filename = '263868__kevcio__amen-break-a-160-bpm.wav'
wave = thinkdsp.read_wave(filename)
wave.normalize()
```

Figure 11-3 (top) shows the spectrum of this wave. Now here's the function that samples from the wave:

```
def sample(wave, factor=4):
    ys = np.zeros(len(wave))
    ys[::factor] = wave.ys[::factor]
    return thinkdsp.Wave(ys, framerate=wave.framerate)
```

I'll use it to select every fourth element:

```
sampled = sample(wave, 4)
```

The result has the same frame rate as the original, but most of the elements are zero. If you play the sampled wave, it doesn't sound very good. The sampling process introduces high-frequency components that were not in the original.

Figure 11-3 (bottom) shows the spectrum of the sampled wave. It contains four copies of the original spectrum (it looks like five copies because one is split between the highest and lowest frequencies).

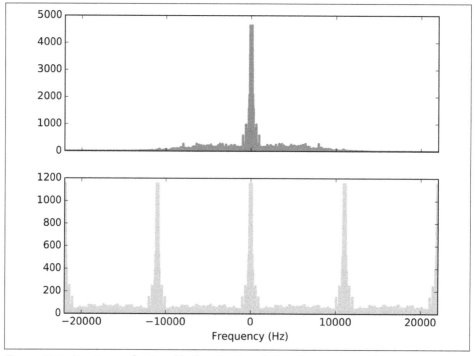

Figure 11-3. Spectrum of a signal before (top) and after (bottom) sampling.

To understand where these copies come from, we can think of the sampling process as multiplication with a series of impulses. Instead of using `sample` to select every fourth element, we could use this function to make a series of impulses, sometimes called an **impulse train**:

```
def make_impulses(wave, factor):
    ys = np.zeros(len(wave))
    ys[::factor] = 1
    ts = np.arange(len(wave)) / wave.framerate
    return thinkdsp.Wave(ys, ts, wave.framerate)
```

And then multiply the original wave by the impulse train:

```
impulses = make_impulses(wave, 4)
sampled = wave * impulses
```

The result is the same; it still doesn't sound very good, but now we understand why. Multiplication in the time domain corresponds to convolution in the frequency domain. When we multiply by an impulse train, we are convolving with the DFT of an impulse train. As it turns out, the DFT of an impulse train is also an impulse train.

Figure 11-4 shows two examples. The top row is the impulse train in the example, with frequency 11,025 Hz. The DFT is a train of four impulses, which is why we get four copies of the spectrum. The bottom row shows an impulse train with a lower frequency, about 5512 Hz. Its DFT is a train of eight impulses. In general, more impulses in the time domain correspond to fewer impulses in the frequency domain.

In summary:

- We can think of sampling as multiplication by an impulse train.
- Multiplying by an impulse train corresponds to convolution with an impulse train in the frequency domain.
- Convolution with an impulse train makes multiple copies of the signal's spectrum.

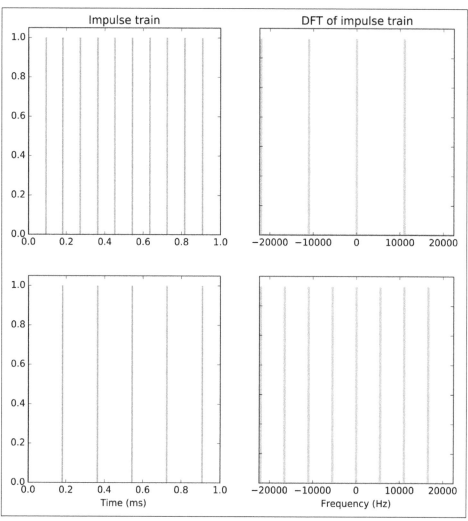

Figure 11-4. The DFT of an impulse train is also an impulse train.

Aliasing

In "Amplitude Modulation" on page 134, after demodulating an AM signal we got rid of the extra copies of the spectrum by applying a low-pass filter. We can do the same thing after sampling, but it turns out not to be a perfect solution.

Figure 11-5 shows why not. The top row is the spectrum of the drum solo. It contains high-frequency components that extend past 10 kHz. When we sample this wave, we convolve the spectrum with the impulse train (second row), which makes copies of the spectrum (third row). The bottom row shows the result after applying a low-pass filter with a cutoff at the folding frequency, 5512 Hz.

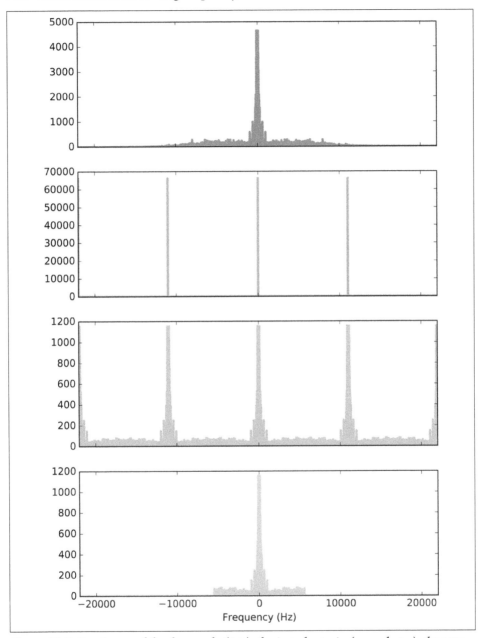

Figure 11-5. Spectrums of the drum solo (top), the impulse train (second row), the sampled wave (third row), and the result after low-pass filtering (bottom).

If we convert the result back to a wave, it is similar to the original wave, but there are two problems:

- Because of the low-pass filter, the components above 5500 Hz have been lost, so the result sounds muted.
- Even the components below 5500 Hz are not quite right, because they include contributions from the spectral copies we tried to filter out.

If the spectral copies overlap after sampling, we lose information about the spectrum and we won't be able to recover it.

But if the copies don't overlap, things work out pretty well. As a second example, I loaded a recording of a bass guitar solo.

Figure 11-6 shows its spectrum (top row), which contains no visible energy above 5512 Hz. The second row shows the spectrum of the sampled wave, and the third row shows the spectrum after the low-pass filter. The amplitude is lower because we've filtered out some of the energy, but the shape of the spectrum is almost exactly what we started with. And if we convert back to a wave, it sounds the same.

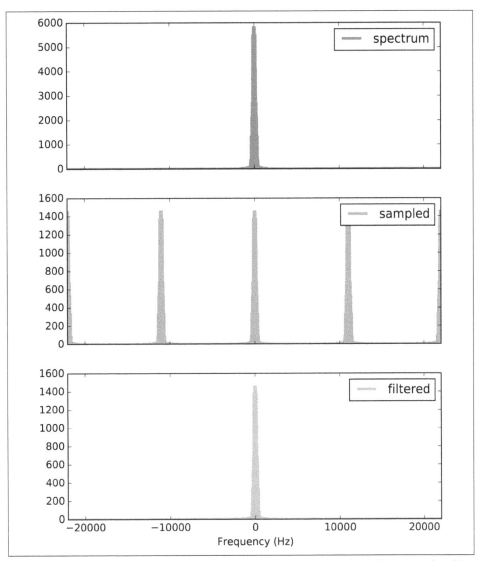

Figure 11-6. Spectrum of a bass guitar solo (top), after sampling (middle), and after fil-tering (bottom).

Interpolation

The low-pass filter I used in the last step is a so-called **brick wall filter**; frequencies above the cutoff are removed completely, as if they hit a brick wall.

Figure 11-7 (right) shows what this filter looks like. Of course, multiplication by this filter in the frequency domain corresponds to convolution with a window in the time domain. We can find out what that window is by computing the inverse DFT of the filter, which is shown in Figure 11-7 (left).

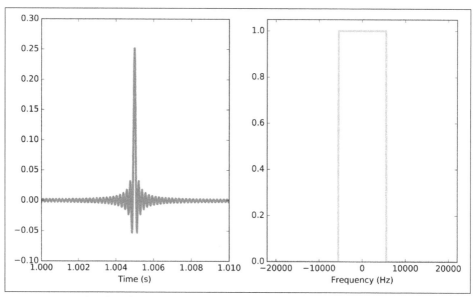

Figure 11-7. A brick wall low-pass filter (right) and the corresponding convolution window (left).

That function has a name—it is the normalized **sinc function**, or at least a discrete approximation of it (see *https://en.wikipedia.org/wiki/Sinc_function*):

$$\text{sinc}(x) = \frac{\sin \pi x}{\pi x}$$

When we apply the low-pass filter, we are convolving with a sinc function. We can think of this convolution as the sum of shifted, scaled copies of the sinc function.

The value of sinc is 1 at 0 and 0 at every other integer value of *x*. When we shift the sinc function, we move the zero point. When we scale it, we change the height at the zero point. So when we add up the shifted, scaled copies, they interpolate between the sampled points.

Figure 11-8 shows how that works using a short segment of the bass guitar solo. The line across the top is the original wave. The vertical gray lines show the sampled values. The thin curves are the shifted, scaled copies of the sinc function. The sum of these sinc functions is identical to the original wave.

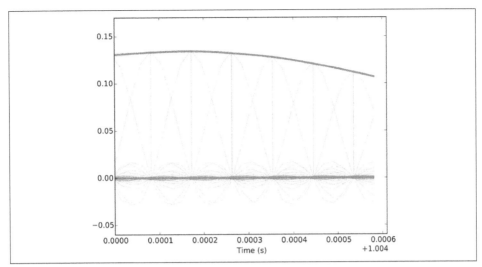

Figure 11-8. Close-up view of a sequence of samples (vertical gray lines), interpolating sinc functions (thin curves), and the original wave (thicker line across the top).

I'll say that again, because it is surprising and important:

> The sum of these sinc functions is identical to the original wave.

Because we started with a signal that contained no energy above 5512 Hz, and we sampled at 11,025 Hz, we were able to recover the original spectrum exactly. And if we have the original spectrum, exactly, we can recover the original wave exactly.

In this example, I started with a wave that had already been sampled at 44,100 Hz, and I resampled it at 11,025 Hz. After resampling, the gap between the spectral copies is 11.025 kHz.

If the original wave contains no energy above 5512 Hz, the spectral copies don't overlap, we don't lose information, and we can recover the original signal exactly.

This result is known as the Nyquist–Shannon Sampling Theorem (see *https://en.wiki pedia.org/wiki/Nyquist-Shannon_sampling_theorem*).

This example does not prove the Sampling Theorem, but I hope it helps you understand what it says and why it works.

Notice that the argument I made does not depend on the original sampling rate, 44.1 kHz. The result would be the same if the original had been sampled at a higher frequency, or even if the original had been a continuous analog signal: if we sample at frame rate f, we can recover the original signal exactly, as long as it contains no energy at frequencies above $f/2$. A signal like that is called **bandwidth limited**.

Summary

Congratulations! You have reached the end of the book (well, except for a few more exercises). Before you close the book, I want to review how we got here:

- We started with periodic signals and their spectrums, and I introduced the key objects in the `thinkdsp` library: `Signal`, `Wave`, and `Spectrum`.

- We looked at the harmonic structure of simple waveforms and recordings of musical instruments, and we saw the effect of aliasing.

- Using spectrograms, we explored chirps and other sounds whose spectrum changes over time.

- We generated and analyzed noise signals, and characterized natural sources of noise.

- We used the autocorrelation function for pitch estimation and additional characterization of noise.

- We learned about the Discrete Cosine Transform (DCT), which is useful for compression and also a step toward understanding the Discrete Fourier Transform (DFT).

- We used complex exponentials to synthesize complex signals, then we inverted the process to develop the DFT. If you did the exercises at the end of Chapter 7, you implemented the Fast Fourier Transform (FFT), one of the most important algorithms of the 20th century.

- Starting with smoothing, I presented the definition of convolution and stated the Convolution Theorem, which relates operations like smoothing in the time domain to filters in the frequency domain.

- We explored differentiation and integration as linear filters, which is the basis of spectral methods for solving differential equations. It also explains some of the effects we saw in previous chapters, like the relationship between white noise and Brownian noise.

- We learned about LTI system theory and used the Convolution Theorem to characterize LTI systems by their impulse response.

- I presented amplitude modulation (AM), which is important in radio communication and also a step toward understanding the Sampling Theorem, a surprising result that is critical for digital signal processing.

If you got this far, you should have a good balance of practical knowledge (how to work with signals and spectrums using computational tools) and theory (an understanding of how and why sampling and filtering work).

I hope you had some fun along the way. Thank you!

Exercises

Solutions to these exercises are in `chap11soln.ipynb`.

Exercise 11-1.

The code in this chapter is in `chap11.ipynb`. Read through it and listen to the examples.

Exercise 11-2.

Chris "Monty" Montgomery has an excellent video called "D/A and A/D | Digital Show and Tell"; it demonstrates the Sampling Theorem in action, and presents lots of other excellent information about sampling. Watch it at *https://www.youtube.com/watch?v=cIQ9IXSUzuM*.

Exercise 11-3.

As we have seen, if you sample a signal at too low a frame rate, frequencies above the folding frequency get aliased. Once that happens, it is no longer possible to filter out these components, because they are indistinguishable from lower frequencies.

It is a good idea to filter out these frequencies *before* sampling; a low-pass filter used for this purpose is called an **anti-aliasing filter**.

Returning to the drum solo example, apply a low-pass filter before sampling, then apply the low-pass filter again to remove the spectral copies introduced by sampling. The result should be identical to the filtered signal.

Index

square waveform, 16, 23, 92
standard deviation, 53
standardizing, 63
static, 40
STFT, 30
stock prices, 116
stretch function, 11
SumSignal, 5
symmetric matrix, 70
synthesis, 11, 65, 66, 80
system, 119

T

temperament, 4
timbre, 3
time domain, 97, 105
time resolution, 31
time-invariant system, 119
timestep, 6
transducer, 1
transfer function, 123, 125, 128
transpose, 70, 85
triangle waveform, 13, 23, 73
trombone, 37
tuning fork, 2

U

UG noise, 49
unbias, 15
uncorrelated noise, 39
uniform noise, 39
unitary matrix, 85
UU noise, 39, 56

V

vector, 70
video, 147
violin, 3, 123
virtual machine, ix
Voss–McCartney algorithm, 51
vowel, 38

W

WAV file, 7, 137
Wave, 5, 9
waveform, 3
white noise, 42, 106, 121
window, 32, 33, 92, 104, 107, 121
windowing, 33

About the Author

Allen B. Downey is a Professor of Computer Science at Olin College of Engineering. He has taught at Wellesley College, Colby College and U.C. Berkeley. He has a Ph.D. in Computer Science from U.C. Berkeley and Master's and Bachelor's degrees from MIT.

Colophon

The animal on the cover of *Think DSP* is a smooth-billed ani (*Crotophaga ani*), a large bird that is part of the cuckoo family. It is found in Florida, the Bahamas, Caribbean islands, and parts of Central and South America.

Smooth-billed anis have black plumage, long tails, and large ridged beaks. They feed on the ground, with a diet made up of termites, insects, and even small lizards and frogs. The birds prefer semi-open habitats with a mix of fields and brushy thickets. As human settlements and deforestation have affected their territory, anis have adapted by frequenting farm pastures and eating the insects flushed out by livestock.

This species is very social and is always found in noisy groups. Mating pairs nest communally with several other couples, taking turns to construct a bowl-shaped nest high in a tree, incubate eggs, and feed the chicks. Each female lays 4–7 eggs, but nests have been found with up to 29 eggs.

Many of the animals on O'Reilly covers are endangered; all of them are important to the world. To learn more about how you can help, go to *animals.oreilly.com*.

The cover image is from the *Braukhaus Lexicon*. The cover fonts are URW Typewriter and Guardian Sans. The text font is Adobe Minion Pro; the heading font is Adobe Myriad Condensed; and the code font is Dalton Maag's Ubuntu Mono.

Have it your way.

Get even more for your money.

Join the O'Reilly Community, and register the O'Reilly books you own. It's free, and you'll get:

- $4.99 ebook upgrade offer
- 40% upgrade offer on O'Reilly print books
- Membership discounts on books and events
- Free lifetime updates to ebooks and videos
- Multiple ebook formats, DRM FREE
- Participation in the O'Reilly community
- Newsletters
- Account management
- 100% Satisfaction Guarantee

Signing up is easy:

1. Go to: oreilly.com/go/register
2. Create an O'Reilly login.
3. Provide your address.
4. Register your books.

Note: English-language books only

To order books online:
oreilly.com/store

For questions about products or an order:
orders@oreilly.com

To sign up to get topic-specific email announcements and/or news about upcoming books, conferences, special offers, and new technologies:
elists@oreilly.com

For technical questions about book content:
booktech@oreilly.com

To submit new book proposals to our editors:
proposals@oreilly.com

O'Reilly books are available in multiple DRM-free ebook formats. For more information:
oreilly.com/ebooks

Milton Keynes UK
Ingram Content Group UK Ltd.
UKHW012338190824
447149UK00007B/120

9 781491 938454